BrightRED Study Guide

Curriculum for Excellence

N5

GEOGRAPHY

Ralph Harnden

First published in 2014 by:
Bright Red Publishing Ltd
1 Torphichen Street
Edinburgh
EH3 8HX

Reprinted with corrections in 2016, 2018, 2021

Copyright © Bright Red Publishing Ltd 2014

Cover image © Caleb Rutherford

All rights reserved. No part of this publication may be reproduced, stored in a retrieval system, or transmitted in any form or by any means, electronic, mechanical, photocopying, recording or otherwise, without prior permission in writing from the publisher.

The right of Ralph Harnden to be identified as the author of this work have been asserted by him in accordance with sections 77 and 78 of the Copyright, Designs and Patents Act 1988.

A CIP record for this book is available from the British Library

ISBN 978-1-906736-38-5

With thanks to:
PDQ Digital Media Solutions Ltd (layout) and Anglosphere Editing Limited (copy-edit)

Cover design and series book design by Caleb Rutherford – e i d e t i c

Acknowledgements
Every effort has been made to seek all copyright holders. If any have been overlooked, then Bright Red Publishing will be delighted to make the necessary arrangements.

Permission has been sought from all relevant copyright holders and Bright Red Publishing are grateful for the use of the following:

Peter Withers (p 8); Ralph Harnden (p 9); Mike Charles/Shutterstock.com (p 11); Carlos Munoz/Shutterstock.com (p 12); Ralph Harnden (pp 17 & 19); Roberto Caucino/Shutterstock.com (p 19); Paul Kennedy (public domain) (p 21); Ralph Harnden (p 21); Hannu (public domain) (p 22); Ralph Harnden (p 23); Stanley Howe (CC BY-SA 2.0)[1] (p 24); Eigenes Werk (public domain) (p 25); Ben Rudiak-Gould (public domain) (p 27); Dchauy/Shutterstock.com (p 28); Ordnance Survey © Crown Copyright. All rights reserved. Licence number 100049324 (pages 31–33, 39 and 60–61); Lindsay Wilson (CC BY 2.0)[2] (p 34); Mark (CC BY-SA 2.0)[1] (p 34); Nick Bramhall (CC BY-SA 2.0)[1] (p 34); ChameleonsEye/Shutterstock.com (p 35); Steve Partridge (CC BY-SA 2.0)[1] (p 36); Route One Publishing Ltd, taken from www.aggbusiness.com (p 37); SAGT (p 37); Lafarge Tarmac Limited (p 37); Richard Penn (CC BY 2.0)[2] (p 38); Richard Penn (CC BY 2.0)[2] (p 38); SSE (p 39); Paul Birrell (CC BY-SA 2.0)[1] (p 39); Atlantis Resources & AR1000 tidal power turbine (p 41); James Cridland (CC BY 2.0)[2] (p 43); leungchopan/Shutterstock.com (p 43); thinboyfatter (CC BY 2.0)[2] (p 44); Patrick Rudolph (CC BY 2.0)[2] (p 45); Lukas Hofstetter (CC BY 2.0)[2] (p 45); gualtiero boffi/Shutterstock.com (p 48); Syda Productions/Shutterstock.com (p 48); Adisa/Shutterstock.com (p 48); Elena Elisseeva/Shutterstock.com (p 48); Giuseppe_R/Shutterstock.com (p 48); Lim Yong Hian/Shutterstock.com (p 48); Elena Stepanova/Shutterstock.com (p 48); Alistair Michael Thomas/Shutterstock.com (p 48); NataliTerr/Shutterstock.com (p 48); Jenny Downing (CC BY-SA 2.0)[1] (p 48); Denis Rozhnovsky/Shutterstock.com (p 48); dp Photography/Shutterstock.com (p 48); Chris McKee (CC BY-ND 2.0)[3] (p 48); Przemysław Sakrajda (CC BY-SA 2.0)[1] (p 48); Wolfman-K (CC BY-SA 2.0)[1] (p 51); areeya_ann/Shutterstock.com (p 51); Julia Ivantsova/Shutterstock.com (p 51); Qpic-Images (p 51); 401(K) 2012 (CC BY-SA 2.0)[1] (p 51); Maxx-Studio/Shutterstock.com (p 51); Ocskay Bence/Shutterstock.com (p 51); albund/Shutterstock.com (p 51); gdvcom/Shutterstock.com (p 52); Colin Smith (CC BY-SA 2.0)[1] (p 58); Ralph Harnden (p 58); Val Vannet (CC BY-SA 2.0)[1] (p 58); Ralph Harnden (p 59); RCAHMS (p 59); Boon Low (CC BY-ND 2.0)[3] (p 62); Ad Meskens (CC BY-SA 3.0)[4] (p 62); Kim Traynor (CC BY-SA 3.0)[4] (p 63); Ralph Harnden (p 64); Danny Fowler (CC BY-SA 2.0)[1] (p 64); Graeme Maclean (CC BY 2.0)[2] (p 64); Ralph Harnden (p 65); Craig Murphy (CC BY 2.0)[2] (p 66); Buccleuch Property (p 67); Blakely074 (CC BY 3.0)[5] (p 67); Pincasso/Shutterstock.com (p 68); chensiyuan (CC BY-SA 3.0)[4] (p 68); Mark Schwettmann/Shutterstock.com (p 68); Sebástian Freire (CC BY-SA 2.0)[1] (p 69); Agência Brasil (CC BY 3.0 BR)[6] (p 69); Colliers Farm Shop (p 70); Foel Farm Park and Anglesey Farmhouse Chocolates (p 70); Colin Smith (CC BY-SA 2.0)[1] (p 71); Stuart Richards (CC BY-ND 2.0)[3] (p 71); Logo © The Soil Association (p 72); Yeo Valley HQ © Soil Association (p 73); Logo © Yeo Valley Farms (Production) Limited (p 73); Paul McIlroy (CC BY-SA 2.0)[1] (p 74); © AgroNotizie - Image Line www.agronotizie.it (p 75); Plantagon/Sweco/Creative Commons (CC BY 2.0)[2] (p 75); International Rice Research Institute (IRRI)/Creative Commons (CC BY 2.0)[2] (p 76); Gideon (CC BY 2.0)[2] (p 77); JasonParis (CC BY 2.0)[2] (p 77); Mariordo (CC BY-SA 3.0)[4] (p 78); Sustainable Sanitation (CC BY 2.0)[2] (p 79); Logo © Practical Action (p 79); R. Weinberg (p 81); Miguel Garces (p 81); American Center Mumbai (CC BY-ND 2.0)[3] (p 83); Spsmiler (public domain) (p 83); Dawn Endico (CC BY-SA 2.0)[1] (p 84); U.S. Fish and Wildlife Service (public domain) (p 85); Judith Doyle (CC BY-ND 2.0)[3] (p 86); U.S. Fish and Wildlife Service (public domain) (p 87); Greg Brookes (p 89); U.S. Fish and Wildlife Service (public domain) (p 89); costaricatroubadour (CC BY 2.0)[2] (p 91); Lou-Bay (CC BY-SA 2.0)[1] (p 91); Brian Gratwicke (CC BY 2.0)[2] (p 91); Horia Varlan (CC BY 2.0)[2] (p 92); crustmania (CC BY 2.0)[2] (p 92); Euyasik (CC BY-SA 3.0)[4] (p 93); Matt Trostle (CC BY 2.0)[2] (p 93); Reforest The Tropics, Inc (p 95); crustmania (CC BY 2.0)[2] (p 95); Rosino (CC BY-SA 2.0)[1] (p 97); U.S. Geological Survey (public domain) (p 98); Mathias Eick EU/ECHO (CC BY-ND 2.0)[3] (p 98); gnuckx (CC BY 2.0)[2] (p 99); U.S. Geological Survey (public domain) (p 99); U.S. Geological Survey (CC BY 2.0)[2] (p 101); CliNKer (CC BY-SA 2.0)[1] (p 101); NASA (p 103); Patrick McFall (CC BY 2.0)[2] (p 105); Daniel Ramirez (CC BY 2.0)[2] (p 106); Airbus S.A.S. (p 107); McDonald's Restaurants Limited (p 107); Airbus S.A.S. (p 107); Shell Trade Marks reproduced by permission of Shell Brands International AG (p 107); Logo © Samsung (p 107); Logo © the Fairtrade Foundation (p 109); Logo © Forest Stewardship Council® (p 109); Marie-Lan Nguyen (CC BY 2.0)[2] (p 110); Antwain (CC BY-SA 3.0)[4] (p 111); Comrogues (CC BY 2.0)[2] (p 112); Peter Hook (CC BY 2.0)[2] (p 112); Tambako The Jaguar (CC BY-ND 2.0)[3] (p 113); Iryna Melnyk/Shutterstock.com (p 115); Change4Life campaign image © Crown Copyright (p 117); Logos for World AIDS day and Act Aware © National Aids Trust (p 119); Tender Mercies Foundation (p 121); UNICEF/NYHQ2005-2336/Mun (p 122); Manu (CC BY-SA 2.0)[1] (p 123).

(CC BY-SA 2.0)[1] http://creativecommons.org/licenses/by-sa/2.0/
(CC BY 2.0)[2] http://creativecommons.org/licenses/by/2.0/
(CC BY-ND 2.0)[3] http://creativecommons.org/licenses/by-nd/2.0/
(CC BY-SA 3.0)[4] http://creativecommons.org/licenses/by-sa/3.0/
(CC BY 3.0)[5] http://creativecommons.org/licenses/by/3.0/
(CC BY 3.0 BR)[6] http://creativecommons.org/licenses/by/3.0/br/deed.en

Printed and bound in the UK by Ashford Colour Ltd.

CONTENTS

BRIGHTRED STUDY GUIDE: NATIONAL 5 GEOGRAPHY

Introduction 4

PHYSICAL ENVIRONMENTS

Weather: latitude and continentality 6
Weather: altitude and aspect 8
Weather: air masses 1 10
Weather: air masses 2 12
Weather: synoptic charts 14
Weather: low-pressure systems (depressions) 16
Weather: high-pressure systems (anticyclones) 18
Landscapes: upland glaciation 20
Landscapes: coasts 1 22
Landscapes: coasts 2 24
Landscapes: upland limestone 26
Landscapes: rivers 28
Landscapes: O.S. map skills 1 30
Landscapes: O.S. map skills 2 32
Land uses: farming and forestry 34
Land uses: industry 36
Land uses: water storage and supply, and hydro-electric power 38
Land uses: renewable energy 40

HUMAN ENVIRONMENTS

Population: global population distribution 1 42
Population: global population distribution 2 44
Population: socio-economic indicators 1 46
Population: socio-economic indicators 2 48
Population: birth and death rates 1 50
Population: birth and death rates 2 52
Population pyramids 54
The demographic transition model 56
Urban geography: urban zones 58
Urban zones on O.S. maps 60
Urban geography: changes in the central business district 62
Urban geography: changes in the inner city 64

Urban geography: changes on the rural–urban (rurban) fringe 66
Urban geography: cities in the developing world .. 68
Agricultural change: farm diversification, government policy and GM crops 70
Agricultural change: organic farming 72
Agricultural change: new technology 74
Agricultural change in the developing world 1 76
Agricultural change in the developing world 2 78

GLOBAL ISSUES

Climate change: causes and effects 80
Climate change: management 82
Tundra regions 84
Tundra: effects of human activity 86
Managing tundra environments 88
Tropical rainforests 90
Tropical rainforests: causes and effects of land degradation 92
Tropical rainforests: solutions to deforestation 94
Environmental hazards: volcanoes, earthquakes and their causes 96
Environmental hazards: volcanoes – effects and management 98
Environmental hazards: earthquakes – effects and management 100
Environmental hazards: tropical storms 102
Environmental hazards: tropical storms – effects and management 104
Trade and globalisation: world trade 106
Trade and globalisation: impact of world trade patterns 108
Tourism: mass tourism 110
Tourism: ecotourism 112
Health: distribution of disease 114
Health: developed world diseases 116
Health: AIDS – a global disease 118
Health: malaria – a developing world disease 120
Health: cholera and kwashiorkor – developing world diseases 122

GLOSSARY 124

BRIGHTRED STUDY GUIDE: NATIONAL 5 GEOGRAPHY

INTRODUCTION

This book has been written to help supplement the course you are being taught in school. It is intended to give you a chance to consolidate your knowledge by covering all the basic points of the course as well as looking at case studies from around the world. Communities in the twenty first century are increasingly interconnected by globalisation, the internet and social media, so geography is more important than ever in helping us to understand other environments, people and cultures. The up-to-date, original and detailed examples described in this book should help your understanding of each topic, give you facts which may be useful in your exam answers and highlight the relevance of what you are studying.

NATIONAL 5 GEOGRAPHY

National 5 Geography is a National Qualifications course which is externally assessed by SQA. The course assessment consists of the exam paper, which accounts for 80% of your overall mark, and the coursework assignment which accounts for 20% of your overall mark.

The National 5 Geography exam is marked out of 80 and is split into three sections. These are:

- **Physical Environments**
- **Human Environments**
- **Global Issues.**

There is very little choice in the first two sections and you will be expected to answer all the questions, except that in the Physical Environments section you will have the choice of answering a question about **either** rivers or upland limestone **or** about coasts or upland glaciated scenery. In the Global Issues section there will be six questions but you will only be asked to answer **two** of them. Your choice will depend on which topics you have studied in class. The contents page will give you an idea of what is covered in each section of the course.

By covering the course fully and building up a detailed set of notes from which you can revise, you should be able to answer the questions in the exam. The best way to do this is by working hard in school throughout the session. The aim of this book is to help you with this process by providing everything you need to know in one place.

COURSEWORK ASSIGNMENT

The coursework assignment is a piece of original geographical research that you will be asked to carry out. It is essential that you use **at least** two research methods to gather your information, such as two different geography fieldwork techniques, map interpretation or internet research. If you choose a topic for which you gather a lot of your information from the internet, you must also ensure that you have used another research method such as analysing a map or taking notes from a book.

It is likely that you will be asked to carry out most of the research for your assignment in your own time as there will be a lot of coursework to get through in class.

While it is perfectly acceptable to choose a topic for which much of your information might be sourced via the internet (i.e. secondary sources), it might be worth considering whether there is a local topic or geographical issue in your area that you could research in person (i.e. primary sources). This would have the advantage of being a completely original topic and also something that is close enough to allow you to carry out your own fieldwork (i.e. independent research). Your teacher should be able to offer you advice about this.

Your topic should be geographical in nature and include explanations of your findings. It should not be purely descriptive. There are some examples of possible topics on page 5.

contd

Examples of coursework assignment topics:

- A sphere of influence study in your local area.
- The effect of new housing developments on local traffic.
- Comparing the environmental quality of different areas of a town.
- Explaining the factors which affect the land uses on a farm.
- Measuring weather conditions and comparing them with forecasts.
- Recording and explaining the flow of water in a stream over a period of time.
- Finding out people's views on a local development such as a wind farm.
- Explaining the reasons for the varying impact of extreme weather on two different locations.
- The impact of deforestation on indigenous people.
- Explaining the population distribution of an island/area/country.

The coursework assignment section of National 5 Geography is marked out of 20 and will be marked externally by SQA. To complete the assignment you will have to answer a number of set questions about the research that you have carried out. You will have one hour to complete these and you will answer them under exam conditions. To help you with this report about your research, you are allowed to prepare **two A4 sheets** beforehand that show information that you have collected and analysed as part of your assignment. This is referred to as **processed information**. In your report you would be expected to refer to this information, **which will also be submitted to SQA along with your answers**. **You must not copy information** directly from these sheets as you will not receive any credit for this. The table below shows what you will be asked to do in the report about your coursework assignment.

The processed information that you take with you when you write up your report should be produced beforehand, based on your research findings. Below are some examples of what your processed information might consist of. You should take in whatever information is most helpful to you in describing and explaining your findings, but this should not include any information which you intend to copy directly into your assignment report. You should try to **explain** your findings as fully as possible to be able to access all of the available marks.

Coursework assignment report

Task	Marks
1. Write down the topic or issue which you have researched.	0
2. Describe two research methods used to collect information about the topic or issue.	6
3. Describe and explain, with reference to processed information, your findings about the topic or issue. Reach a well-supported conclusion about the topic or issue.	14

Examples of processed information for the coursework assignment report:

- An annotated photograph(s) or field sketch(s).
- An annotated cross-section or land use map.
- A table(s) or graph(s) showing figures about your topic or issue.
- Results of a questionnaire you used to survey shoppers about traffic problems, together with details of the number of people surveyed.
- A letter, with questions, you sent to the local council planning department.
- Data you have gathered from internet sources.
- A screenshot of a list of websites you found using a search engine with the most relevant ones highlighted.
- Weather data and weather maps you have gathered over a certain period of time.
- A spider diagram showing key information taken from watching a video/TV programme/DVD, upon which you can elaborate.
- A mind map containing prompts to help you write your report.

In the report conclusion you should do one or more of the following:

- make decision(s) or recommendation(s) about your topic or issue
- explain the most important things you found out in your research
- provide answers to the question or questions you set yourself at the start of your research
- say whether or not the statement you started with has been proved or disproved.

An early start to your coursework assignment would allow you time to gather plenty of information. If possible, do some of this during the summer holiday when good weather might allow you to carry out physical fieldwork in more pleasant conditions! A good coursework assignment should enable you to secure a high mark in the report, giving you a better chance of a good overall grade.

PHYSICAL ENVIRONMENTS

WEATHER: LATITUDE AND CONTINENTALITY

WEATHER: AN INTRODUCTION

The British are famous for their national trait of talking about the weather. People from other countries are often amused and baffled by this obsession of ours. However, there is good reason for our fascination with the weather. Unlike many other parts of the world, where the weather is quite predictable and often stays the same for many days or even weeks at a time (such as around the Mediterranean in summer), the weather in the British Isles is very changeable and notoriously difficult to predict. This is due to our location in a particularly turbulent area of the atmosphere where warm air from the Tropics meets colder air arriving from the Arctic. Our position on the edge of the Atlantic Ocean, where powerful weather systems can form and then sweep in over our islands, also results in frequent and sudden changes in weather conditions.

The day-to-day changes in our atmosphere are what we describe as the **weather**. There are eight main **weather elements** and some basic knowledge of these is helpful in understanding the weather (see table below). There are many factors which affect our weather and in particular the temperature. Some of these factors are explained on the next few pages.

Weather elements

Weather element	Units of measurement	Weather instrument
Temperature	Degrees centigrade	Maximum–minimum thermometer
Sunshine	Hours	Sunshine recorder
Precipitation	Millimetres	Rain gauge
Wind speed	Kilometres/miles per hour	**Anemometer**
Wind direction	Points of the compass	Wind vane
Air pressure	Millibars	Barometer
Cloud cover	Oktas	Estimate
Visibility	Kilometres	Estimate

ONLINE
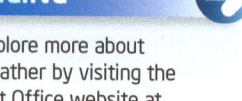
Explore more about weather by visiting the Met Office website at www.brightredbooks.net/N5Geography

DON'T FORGET

The Sun's rays are more spread out the further north of the equator you go, therefore the heating effect is lower and so are the average temperatures.

LATITUDE

Diagram 1.1 Average summer temperatures affected by latitude.

All weather patterns on Earth are caused by the Sun. Energy from the Sun that arrives in our atmosphere is known as **insolation**. The amount of insolation received on each part of the planet's surface depends on its distance from the equator. How far north or south of the equator a place is can be measured by its latitude. The North Pole, for example, is 90° north (90°N) of the equator while Edinburgh is 55°N. Generally, the further away from the equator a place is the less insolation it will receive, so it is likely to be cooler (Diagram 1.1).

This change in the amount of insolation is due to the curvature of the Earth's surface. Where the Sun is almost directly overhead, in the Tropics, its rays are focused intensely on the planet surface, concentrating the Sun's energy and making it hotter. Where the Sun is lower in the sky, in Britain for example, the rays strike the Earth's surface at a lower angle and so spread out more. This makes them less concentrated and results in cooler temperatures (see Diagram 1.2).

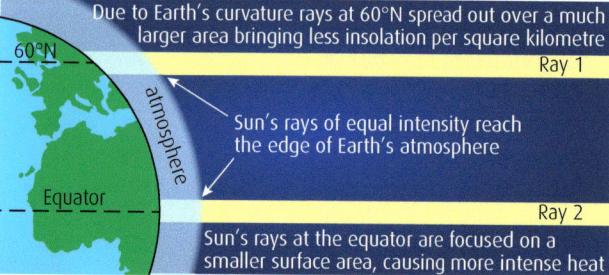

Diagram 1.2 How latitude and the Earth's curvature affect insolation.

CONTINENTALITY

Continentality is the effect which large land masses have on the weather. In particular, continentality has an effect on temperatures, with areas far inland, away from the influence of the sea, having more extreme temperatures. Even within Scotland it is common for places away from the sea, such as Aviemore, to have colder temperatures in winter and hotter temperatures in summer than places on the coast. This is due to the different heat capacities of land and sea. The sea heats up very slowly but retains its heat for months, whereas the land heats up quickly but then loses it again quickly, within hours. Places close to the sea therefore benefit from some of its stored heat energy during the winter months.

This means that in places next to the sea such as Britain, temperatures are cooler in summer and warmer in winter than places further inland but at the same **latitude**, such as Russia. Because of continentality, Edinburgh (55°N), for example, has a January average of 3°C and a July average of 15°C, while Moscow (55°N), which is further away from the sea, has a January average of –8°C and a July average of 17°C. Also at the same latitude, the Russian city of Kazan (55°N), which is further inland still, has a January average of –13°C and a July average of 20°C. Being close to the sea usually means temperatures are less extreme. Diagram 1.3 shows the effects of continentality. It is important to point out that temperatures in Britain are kept higher than average due to a warm sea current, the **North Atlantic Drift**, which brings stored heat from the Tropics as it flows past the British Isles.

Diagram 1.3 Average temperatures affected by continentality.

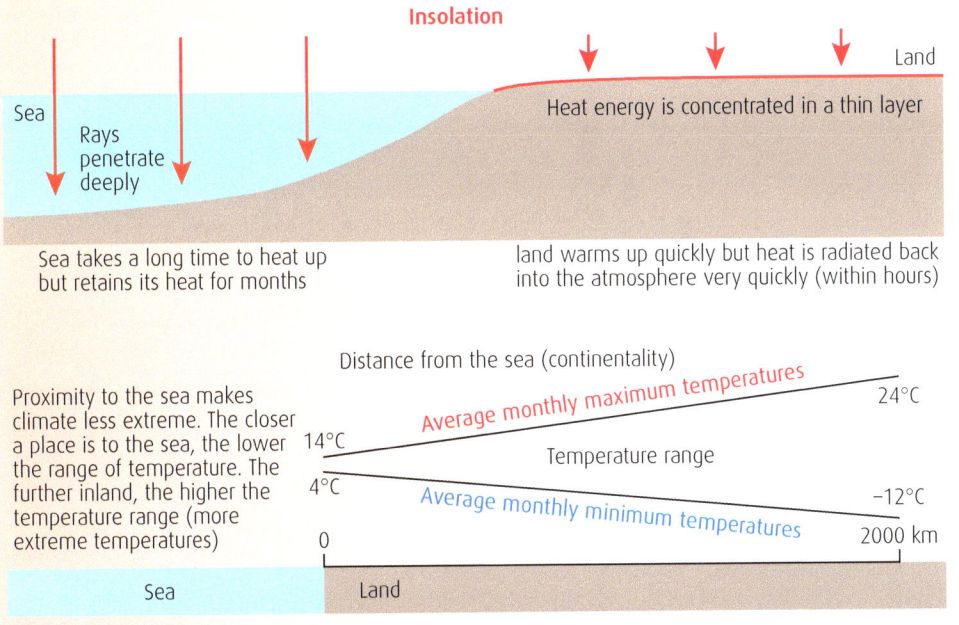

Diagram 1.4 Reasons for continentality.

> **DON'T FORGET**
>
> Temperatures in Britain are kept higher than average for its latitude in winter due to the warming effect of the North Atlantic Drift.

> **DON'T FORGET**
>
> The further inland you go, the more extreme the temperatures become: warmer in summer and cooler in winter. This is because the sea acts as a thermal reservoir in winter and keeps the temperatures of coastal areas up. In summer, coastal areas are kept cooler by sea breezes because the sea doesn't heat up as intensely as the land (Diagram 1.4).

> **VIDEO LINK**
>
> Learn more about the Earth's climate system by watching the video 'How does the climate system work?' at www.brightredbooks.net/N5Geography

THINGS TO DO AND THINK ABOUT

1 **Explain** why average temperatures are usually colder the further away you go from the equator.
2 Make a star diagram to show eight different weather elements and their units of measurement.

> **ONLINE TEST**
>
> Visit the BrightRED Digital Zone (www.brightredbooks.net/N5Geography) and take the 'Latitude and Continentality' test to revise this topic.

PHYSICAL ENVIRONMENTS
WEATHER: ALTITUDE AND ASPECT

ALTITUDE

Altitude is the height of a place above sea level. As altitude increases, the average temperature drops. The rate of temperature decrease depends on the weather, with the fall being greater in clear, cloudless conditions. The rate of this temperature decrease is approximately 6.5°C for every 1000 metres above sea level. Why does this happen?

The air is heated indirectly by the Sun. As the Sun's rays enter the Earth's atmosphere they mostly pass through it as short-wave radiation without heating the air. When the Sun's rays reach the Earth's surface, the heat energy is absorbed by the land (or sea) and reradiated back as long-wave radiation, which then heats the air. The air is also denser at sea level than at higher altitudes, so it gets warmer more quickly. Most of the time, therefore, the places closest to sea level are warmest while places that are higher have lower average temperatures. For example, Braemar (339 metres above sea level) has an average July maximum of 18°C, while the average July maximum for Paisley (32 metres above sea level) is 20°C. This is why it can be cold enough for snow at the top of a mountain but not at the bottom (see Diagram 1.5).

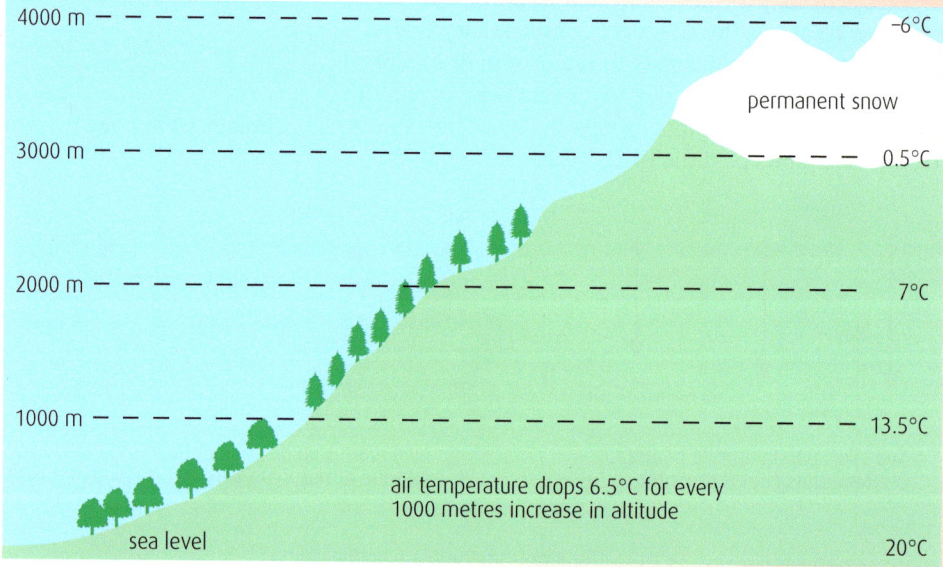

Diagram 1.5 Average temperatures affected by altitude (example from Swiss Alps).

DON'T FORGET

At higher altitudes there are fewer air molecules than at sea level, where they are more densely packed. High places are, therefore, usually cooler because there is less air to absorb the Sun's heat.

The south-facing slopes of the Upper Rhone Valley above Sion in Switzerland show the changing vegetation from vineyards in the warm climate near the valley bottom, to coniferous forest and finally small Alpine plants and bare rock in the much colder climate at the mountain tops. This is evidence of how average temperature falls with increasing height above sea level. In this example, farmers are also making use of the favourable south-facing aspect of the valley slopes, which help to increase the average temperature and make it possible to grow vines at higher altitudes.

Physical Environments – Weather: Altitude and aspect

ASPECT

Weather conditions, especially temperature, can be affected by the direction which a place faces. For example, in hilly areas where a valley runs from east to west, the slopes on the northern side of the valley face south and receive more sunshine and have warmer temperatures than the slopes on the southern side, which face north. The direction in which a place faces is known as its **aspect**. The temperature difference between the north- and south-facing slopes of a valley can be quite dramatic. This is why, for example, corries are often found on the northern side of a mountain, where there has been less sunshine and the snow has been able to lie for longer.

In the Alps it can be warm enough to grow vines on the south-facing slopes, while the north-facing slopes are colder and covered in coniferous forest. Aspect has a big effect on the location of settlements, with sunnier south-facing slopes being popular sites for farms and villages, while the colder north-facing slopes are left as pastures or planted with coniferous trees. The temperature differences between opposite sides of a valley can be exaggerated because of the way in which the surrounding hills provide shelter (see photo opposite).

ONLINE

For more on aspect follow the link at www.brightredbooks.net/N5Geography

Alpine valley: sunny south-facing, shaded north-facing slopes.

SUMMARY: FACTORS AFFECTING TEMPERATURE

The average temperature of a place is influenced by:

- **latitude**: on average, temperatures decrease towards the poles
- **continentality**: on average, places far inland have more extreme temperatures than places close to the sea
- **altitude**: on average, temperatures fall with increasing height above sea level
- **aspect**: on average, places in the northern hemisphere which face south have higher temperatures than places which face north.

DON'T FORGET

The type of air mass moving over a place will also influence the temperature at that time. Different air masses bring different temperature conditions.

ONLINE TEST

Take the 'Altitude and aspect' test online at www.brightredbooks.net/N5Geography

THINGS TO DO AND THINK ABOUT

1. Explain why places that are higher have lower average temperatures.
2. An estate agent selling a house with a **southerly aspect** highlights this as an important selling point to potential buyers. Explain how a southerly aspect could benefit occupants of the house.

PHYSICAL ENVIRONMENTS
WEATHER: AIR MASSES 1

AIR MASSES: AN OVERVIEW

An **air mass** is a large body or chunk of air, the weather characteristics of which are defined by the area where it has developed and from where it is coming. An air mass can be many hundreds of kilometres across, perhaps even a thousand or more. There are five different air masses that affect the British Isles (see diagram). Each air mass brings its own weather and can sometimes affect us for many weeks. Because different air masses frequently meet over the British Isles, weather **fronts** are formed (where the edges of different air masses meet), bringing their own particular sort of weather.

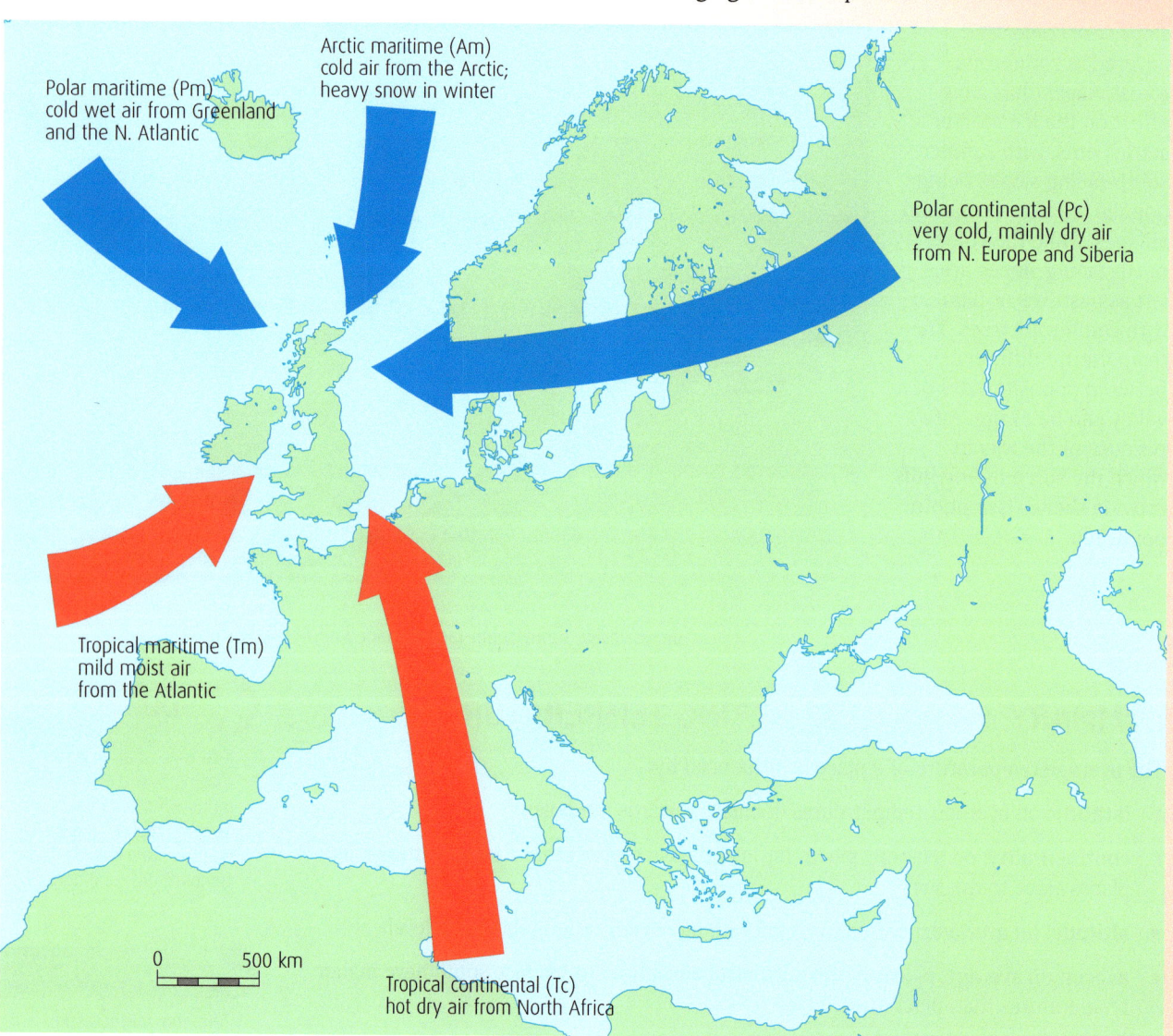

Diagram 1.6 Air masses affecting the British Isles.

VIDEO LINK

For more, check out the clip 'What are air masses?' at www.brightredbooks.net/N5Geography

Air masses can be tropical, polar or Arctic. A **tropical** air mass develops near the Tropics and brings mild or warm air. A **polar** air mass develops further north and brings colder weather. An **Arctic** air mass moves down across Britain from north of the Arctic Circle and can bring exceptionally cold weather. Air masses can also be continental or maritime. A continental air mass develops over the land and can bring quite dry weather. A maritime air mass develops over the ocean and usually brings moist air, causing showery, wet conditions.

TROPICAL CONTINENTAL

A **tropical continental** air mass forms over land in or close to the Tropics. Consequently, these air masses are usually warm and dry. For example, a tropical continental air mass might form over north Africa and although it will pass over the Mediterranean on its journey to the British Isles, it will still be mainly dry and bring with it the warm temperatures associated with that part of the world. A tropical continental air mass in summer might bring a prolonged dry spell and a heat wave, especially to the more southern parts of the British Isles. In 1976 a tropical continental air mass was responsible for a severe **drought** in the UK, resulting in widespread water use restrictions such as hosepipe bans. It also caused hardship to farmers when grass withered and died back, resulting in loss of grazing land for livestock. Temperatures were so high that some, especially more elderly, people suffered from and even died of heat stroke. However, many people still remember the continuously sunny, clear conditions and the weeks of gloriously warm, dry weather.

Tropical continental air can bring a heat wave in summer.

TROPICAL MARITIME

A **tropical maritime** air mass forms over warm ocean areas to the south of the British Isles, therefore they generally bring mild and wet weather. There is a lot of moisture in this air mass that is evaporated from the oceans, the air is humid and as it is forced to rise over the land, it cools and condenses, forming clouds and sometimes rain. Tropical maritime air masses are blown in by prevailing south-westerly winds and are the most common type of air mass to affect Britain. The warm rising air can create areas of low pressure, which cause further rain, so these air masses bring changeable conditions. The term 'mild' should be used with caution because mild air may simply mean that it is a few degrees warmer than the air that it replaces – in winter this could still be quite cold!

DON'T FORGET

Although the general weather conditions are determined by the source of the air affecting the country, the precise weather conditions in any one place will still be affected by other factors such as latitude, aspect, altitude and distance from the sea.

ONLINE

Learn more about air masses by following the link at the BrightRED Digital Zone.

ONLINE TEST

Take the 'Air masses 1' test online at www.brightredbooks.net/N5Geography

THINGS TO DO AND THINK ABOUT

1. Why does Britain's weather change so frequently?
2. A heat wave forecast for the UK is good news. Do you agree? Give reasons to support your answer.

PHYSICAL ENVIRONMENTS

WEATHER: AIR MASSES 2

POLAR CONTINENTAL

A **polar continental** air mass forms over land to the north and east of the British Isles. The main characteristic of this air mass is the cold air that it often brings in winter. Although polar continental air masses are quite dry, they pick up moisture as they cross the North Sea, so rain showers, **haar** and, in winter, snowfalls are often experienced as a result, especially along the east coast. Polar continental air masses can often affect the British Isles as a result of blocking high-pressure systems over Scandinavia in winter, drawing in particularly cold air from Russia and Siberia along their southern edge. This was the cause of the particularly cold and prolonged wintry weather conditions in 2010, when the UK's lowest ever November temperature of −18°C was recorded in Powys, mid-Wales. A similar blocking high-pressure system over Scandinavia and the resulting polar continental air mass in March and April 2013 saw the coldest temperature ever recorded on Easter Day in the UK: −12.5°C in Braemar on 31 March.

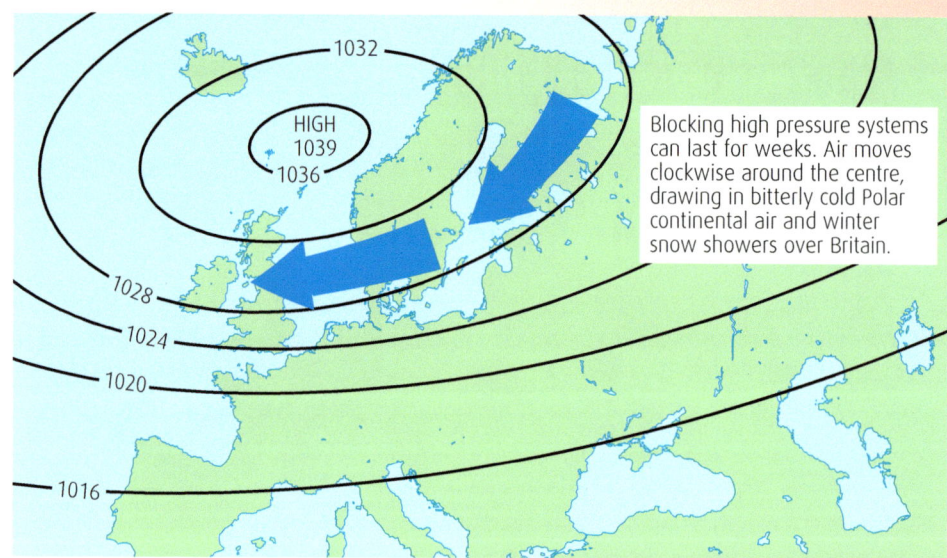

Blocking high pressure systems can last for weeks. Air moves clockwise around the centre, drawing in bitterly cold Polar continental air and winter snow showers over Britain.

Diagram 1.7 A blocking high-pressure system in winter, drawing a polar continental air mass towards Britain.

VIDEO LINK

Check out the clip about record low temperatures in the UK in March 2013 at www.brightredbooks.net/N5Geography

DON'T FORGET

Once an **air mass** has moved out of the area in which it formed it is called an **air stream**. The type of weather that an air stream brings to Britain depends on the area it started from and the areas over which it passes before arriving in the British Isles.

Polar continental air in winter can bring snow as well as freezing conditions.

POLAR MARITIME

Polar maritime air masses form over the cold areas of the Atlantic Ocean to the north and west of the British Isles. As with tropical maritime air masses, there is a great deal of moisture evaporated from the surface of the ocean, so the weather conditions associated with polar maritime air masses are cool and wet. Showery weather is very common as air rises up on meeting the land, and air temperatures are depressed as the wind arrives in the British Isles having passed over the cooler waters of the North Atlantic. Polar maritime air masses are often experienced after a **depression** passes over the UK and the wind draws in cool, moist air from a north-westerly direction.

ARCTIC MARITIME

An **Arctic maritime** air mass is, as its name suggests, a profoundly cold air mass formed north of the Arctic Circle, often over the ice fields of Greenland or over the sea pack ice towards the North Pole. In winter an Arctic maritime air mass brings a real threat of snow and wintry conditions. Temperatures are very low, often below zero, and because the air passes over open ocean to the north of the British Isles it is humid. This combination can bring heavy snowfalls and blizzard conditions, especially along north-facing coasts, but a large Arctic maritime air mass can plunge the whole of the UK into Arctic conditions for many days. Scotland's ski centres may positively relish the arrival of an Arctic maritime air mass at the beginning of winter, as their ski slopes will be well covered with snow and ready for business!

SUMMARY

- An **air mass** is a huge body of air with particular weather characteristics.
- When an air mass moves from its source area it becomes an **air stream**.
- **Tropical continental** air masses bring warm or hot dry weather in summer. This can sometimes set off convectional thunderstorms.
- **Tropical maritime** air masses bring mild moist air, often causing cloudy and wet conditions.
- **Polar continental** air masses bring very cold weather in winter, often with snow showers.
- **Polar maritime** air masses bring cool and wet weather.
- **Arctic maritime** air masses bring bitterly cold weather and significant snowfalls in winter.

ONLINE

For more on air masses affecting the UK follow the link at www.brightredbooks.net/N5Geography

ONLINE TEST

Test yourself on 'Air masses 2' online at www.brightredbooks.net/N5Geography

THINGS TO DO AND THINK ABOUT

1. List the five main air masses that affect the British Isles and say what type of weather conditions they bring.
2. For each of the following air masses give both the **advantages** and **problems** which they could cause for people in the UK.
   (a) A **tropical continental** air mass in **summer**.
   (b) A **polar continental** air mass in **winter**.

PHYSICAL ENVIRONMENTS
WEATHER: SYNOPTIC CHARTS

SYNOPTIC CHARTS: AN OVERVIEW

A weather map or **synoptic chart** shows the weather patterns over a certain area at a given time (see diagram below). A series of synoptic charts shows how the weather changes over time. These are vital to mariners and aircraft pilots to help plan their journeys safely. Simplified synoptic charts are often used to deliver TV weather forecasts. With an understanding of synoptic charts and the symbols on them, it is possible to make simple predictions about weather conditions in the immediate future.

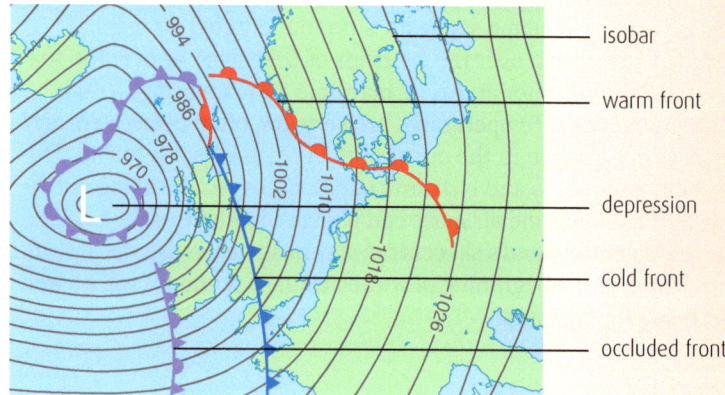

Diagram 1.8 A synoptic chart.

VIDEO LINK
Check out the 'How to read a synoptic chart' clip at www.brightredbooks.net/N5Geography

SYNOPTIC CHART SYMBOLS

To fully understand synoptic charts it is necessary to know the meanings of the various symbols used by the Meteorological Office to show observations recorded at weather stations. These are shown on the diagrams on these two pages. The different types of pressure system which are shown on synoptic charts are explained on pages 16–19.

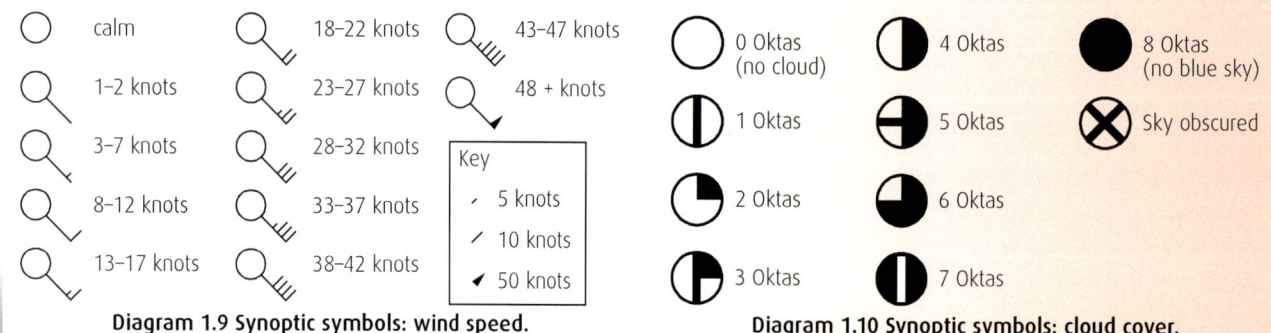

Diagram 1.9 Synoptic symbols: wind speed.

Diagram 1.10 Synoptic symbols: cloud cover.

DON'T FORGET
Wind is always named after the direction **from** which it blows. A north wind comes from the north, bringing cool air as it moves south towards the British Isles. The symbols in Diagram 1.9 all show a south-easterly wind.

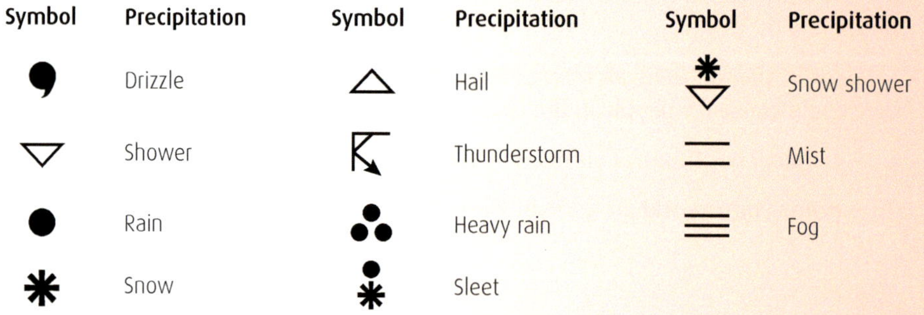

Diagram 1.11 Synoptic symbols: precipitation.

contd

14

Physical Environments – Weather: Synoptic charts

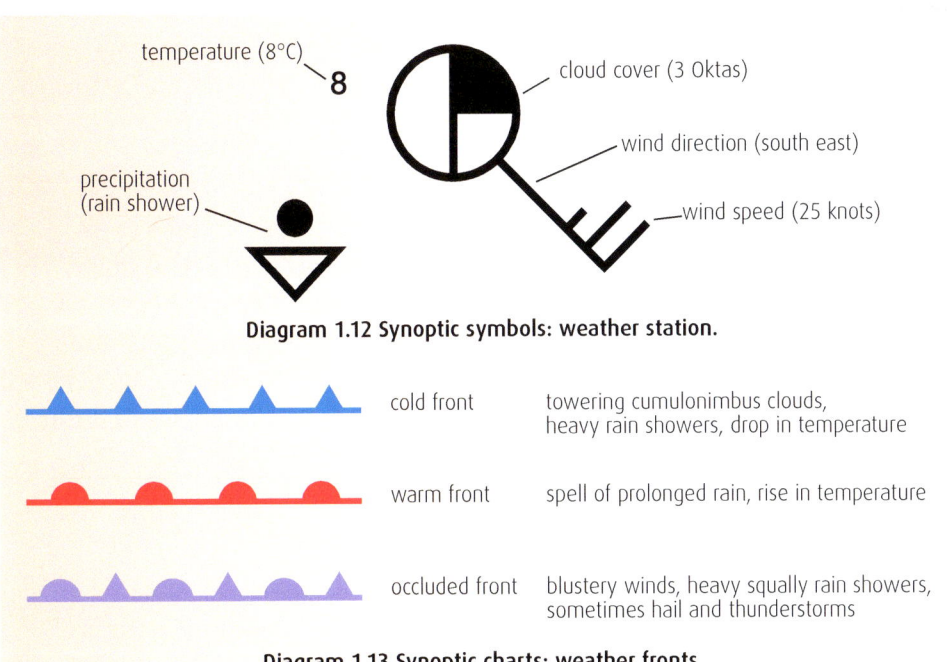

Diagram 1.12 Synoptic symbols: weather station.

▲▲▲▲▲	cold front	towering cumulonimbus clouds, heavy rain showers, drop in temperature
●●●●●	warm front	spell of prolonged rain, rise in temperature
▲●▲●▲●	occluded front	blustery winds, heavy squally rain showers, sometimes hail and thunderstorms

Diagram 1.13 Synoptic charts: weather fronts.

DON'T FORGET
Learn how to draw and interpret synoptic symbols. You will be expected to know how to do this without a key to help you.

ONLINE
Learn more about synoptic charts by following the link at www.brightredbooks.net/N5Geography

ONLINE TEST
Test yourself on synoptic charts at www.brightredbooks.net/N5Geography

THINGS TO DO AND THINK ABOUT

1. For further information about synoptic charts and weather, explore the Meteorological Office website at http://www.metoffice.gov.uk/learning.

2. Regular study of the current synoptic charts for the UK will improve and consolidate your knowledge of weather and weather forecasting. Go to http://www.metoffice.gov.uk/public/weather/surface-pressure.

3. Match these synoptic symbols with the descriptions at the weather stations below.

   (a)   (b)   (c)   (d)   (e)

   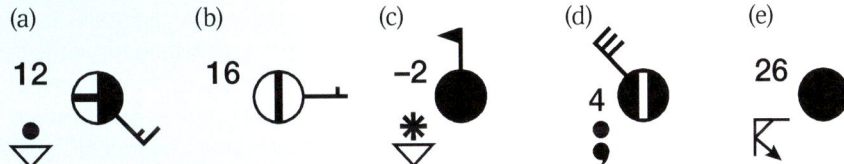

   Altnaharra: Overcast, very cold, strong northerly winds and wintry showers.

   Tiree: Mainly overcast, feeling cold in the strong north-westerly winds, which will bring rain and drizzle.

   Leuchars: Patchy cloud and sunny intervals, feeling mild with mainly light south-easterly winds, which will bring occasional showers.

   Dover: Feeling very humid and muggy in the hot weather, which is likely to set off some electrical storms; mainly calm.

   Pershore: A pleasant, mild and mostly clear day with a very light easterly breeze; staying dry.

4. Draw synoptic symbols to show the following weather conditions:

   (a) 2 oktas, 21°C, dry with 25 knot southerly winds

   (b) 6 oktas, 3°C, hail showers, 40 knot westerly wind

   (c) 7 oktas, 10°C, heavy rain, 30 knot north-easterly wind

   (d) sky obscured, 8°C, fog, 10 knot south-easterly wind

   (e) 8 oktas, −2°C, snow, 35 knot north-westerly wind.

PHYSICAL ENVIRONMENTS

WEATHER: LOW PRESSURE SYSTEMS (DEPRESSIONS)

To achieve a basic understanding of weather patterns it is necessary to study changes in air pressure. Air pressure (measured in millibars) is simply the weight of air that is pressing down on the surface of the planet. Gravity keeps air close to the surface, so generally air pressure is greatest at sea level and gets less with increasing altitude. However, air pressure is constantly changing as a result of specific atmospheric conditions and so areas of high pressure (anticyclones) and low pressure (depressions) are formed. Air tries to move from high pressure zones to areas of low pressure, creating wind. The bigger the difference in air pressure the stronger the wind.

LOW PRESSURE SYSTEMS: AN OVERVIEW

Low air pressure is formed when air rises. This can happen on a hot day, when warm air starts to rise. Convection currents, or thermals, are created: the rising air cools and condenses, and clouds form. If the convection is strong enough, towering **cumulonimbus** clouds are formed and a thunderstorm can occur. Although this can happen, particularly on warm summer days in the British Isles, rising air is more commonly caused by the meeting of different air masses. The point along which two air masses meet is called a front. The warmer, lighter air is forced to rise over the colder, denser air and so low pressure (a depression) starts to form. If there is a particularly big difference in air temperature between the two air masses, uplift can be quite intense and so a deep area of low pressure is formed.

FRONTS

VIDEO LINK

Check out the clip 'Weather fronts' for more at www.brightredbooks.net/N5Geography

DON'T FORGET

When the wind changes direction it can be said to **veer**. The wind **veers** from south-westerly to westerly.

When a depression forms (Diagram 1.14), a warm air mass pushes in to a colder air mass, forming a bulge. As this bulge becomes more pronounced, the leading edge of the warm air creates a **warm front**, while the trailing edge, where it is being pushed forwards by the cold air, is called the **cold front** (see diagram below). Where the cold front catches up with the warm front, pushing the warm air off the ground altogether, an **occluded front** (or occlusion) is formed. Each front brings a particular type of weather with it and so each time a depression passes over a place, a similar pattern of weather occurs. It is helpful to know what this pattern is to be able to accurately forecast the weather.

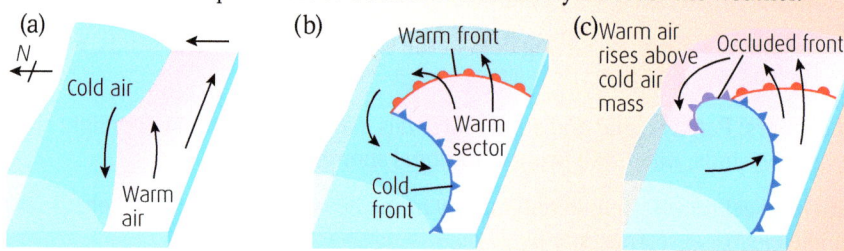

Diagram 1.14 Formation of a depression.

DEPRESSIONS ON SYNOPTIC CHARTS

DON'T FORGET

Learn the symbol for each type of front and also the particular weather it brings.

A depression appears on a synoptic chart as a series of tightly packed concentric pressure lines or **isobars**. Isobars are usually shown at intervals of 4 or 8 **millibars**. The closer together the isobars, the stronger the wind speeds. The relative positions of the warm and cold air masses are marked by fronts. The front edge of the warm air is called the warm front while the trailing edge is the cold front, behind which the cold air mass is located. The area between the two fronts is known as the **warm sector**. A particular type of weather is associated with each part of the depression.

contd

Physical Environments – Weather: Low pressure systems (depressions)

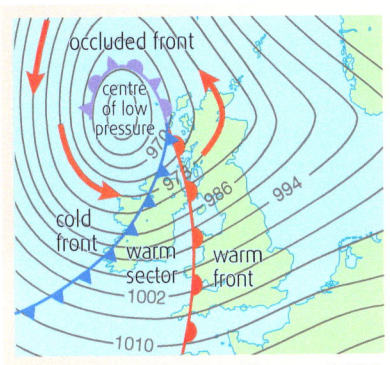

- winds blow anti-clockwise around centre
- isobar values go down towards centre
- tightly packed isobars give strong winds
- fronts give rapidly changing weather conditions

Diagram 1.15 Synoptic chart showing a depression.

DON'T FORGET

Wind blows in an anticlockwise direction around the centre of a depression.

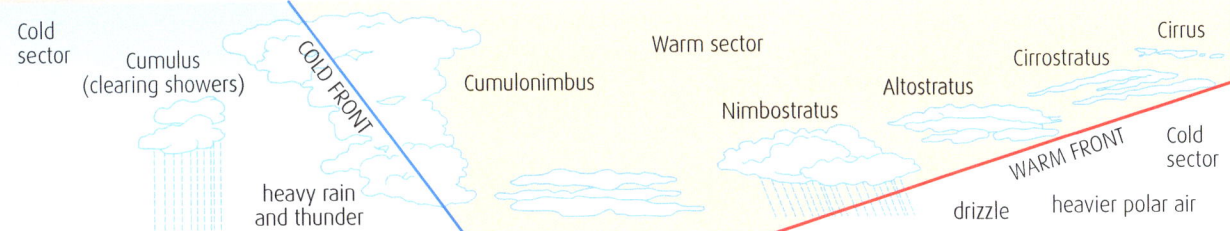

Diagram 1.16 Cross-section through a depression.

	After cold front	At cold front	In warm sector	At warm front	As warm front approaches	Well ahead of warm front
Temperature	3°C	3°C	10°C	7°C	4°C	5°C
Clouds	cumulus	cumulonimbus	broken stratus	nimbostratus	altostratus	cirrus
Precipitation	showers clearing	heavy showers	isolated showers	prolonged rain	drizzle	dry
Wind direction	north west	north west	west	south west	southerly	light southerly
Air pressure	rising	rising	low	falling	falling	high

Weather patterns associated with a depression

WEATHER ASSOCIATED WITH DEPRESSIONS

As a depression approaches, the first sign is often high-level **cirrus** cloud (or grey mares' tails). This is caused by the warm air rising over the denser air of the cold sector. As the front moves closer, there is more uplift and condensation, causing lower and thicker clouds. Drizzle or light rain may begin. By the time the warm front arrives, a thick layer of low-level **nimbostratus** cloud will have formed, giving a prolonged spell of rain. As the warm front passes, the warm sector arrives, giving a marked increase in temperatures. The cloud may also break up, as there is less uplift, and sunny intervals with showers may occur. As the cold front approaches, there is strong uplift as the warm air is pushed up by the colder denser air. This causes towering cumulonimbus clouds to form, giving short, heavy rain showers. As the cold front passes, the temperature drops and the cloud breaks up, leaving clearing skies with scattered showers. Often the wind is strong as the cold front passes and it **veers** from westerly to north-westerly. This pattern of weather is repeated each time a depression moves across the British Isles from the Atlantic towards the North Sea. In areas where an occluded front is experienced, the weather is often very poor, with strong winds, thick cloud and heavy downpours, sometimes with hail and/or thunderstorms.

ONLINE

Learn more about different cloud formations by following the link at www.brightredbooks.net/N5Geography

Cirrus clouds (grey mares' tails) signalling the approach of a weather front.

ONLINE TEST

Test yourself on low pressure systems at www.brightredbooks.net/N5Geography.

THINGS TO DO AND THINK ABOUT

1. Watch regular TV forecasts on programmes such as BBC's Reporting Scotland and try to understand the weather patterns that appear.
2. Learn to become more aware of the clouds in the sky. Sometimes you can spot the tell-tale signs of an approaching front as the wispy cirrus and altostratus clouds begin to appear.

PHYSICAL ENVIRONMENTS

WEATHER: HIGH PRESSURE SYSTEMS (ANTICYCLONES)

HIGH PRESSURE SYSTEMS: AN OVERVIEW

High pressure systems, also known as **anticyclones**, form when there is no uplift, so there is less chance of clouds forming. Anticyclones consist of just one air mass, so there are no fronts and the weather is much less changeable than in a depression. In fact, high pressure systems are known for their stable weather conditions, which can often last for many days or even weeks. The type of weather associated with high pressure depends on the season, but usually conditions are relatively clear and there is not too much wind.

ANTICYCLONES ON SYNOPTIC CHARTS

High pressure appears as a zone of stable weather conditions on a synoptic chart. This means there are no fronts and the **isobars** are much further apart than in a depression. The isobars are more elliptical (oval shaped), unlike the circular isobars in a depression, while the isobar values rise towards the centre. A high pressure zone often covers a large area and can be many hundreds of kilometres across, covering the whole of Scandinavia, for example.

DON'T FORGET

Winds travel in a clockwise direction around high-pressure zones, but because the isobars are widely spaced, winds are usually light.

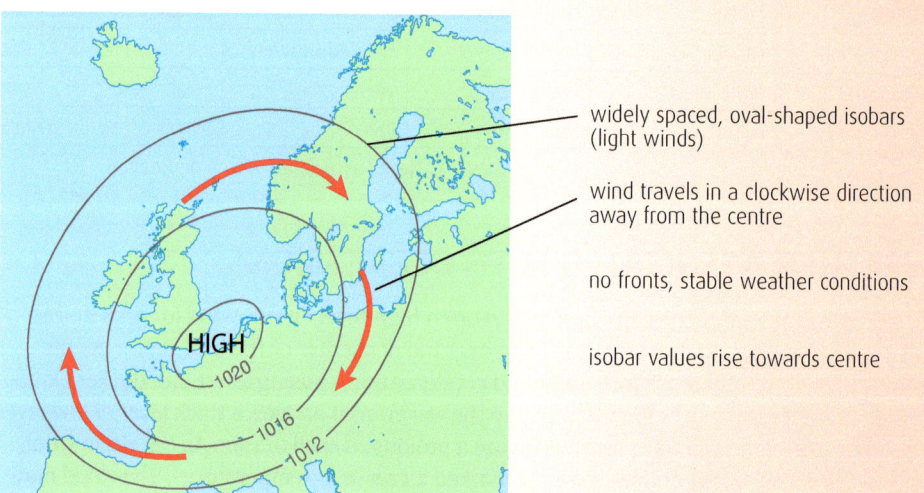

- widely spaced, oval-shaped isobars (light winds)
- wind travels in a clockwise direction away from the centre
- no fronts, stable weather conditions
- isobar values rise towards centre

Diagram 1.17 Synoptic chart showing an anticyclone.

WEATHER ASSOCIATED WITH ANTICYCLONES

Anticyclones bring stable weather conditions. Air is descending within a high pressure zone so there is no uplift, giving fewer clouds and so rain is unusual.

Widely-spaced isobars result in weaker winds, so anticyclones are characterised by light winds or calm, tranquil weather.

Anticyclones in summer and winter bring very different weather conditions.

In summer, the lack of cloud cover allows intense sunshine to reach the land surface, giving high temperatures. High pressure in summer is associated with calm, clear, sunny, hot and dry weather. This can result in a heat wave and, if it lasts long enough, can result in droughts and water restrictions.

In winter, the lack of cloud cover also gives bright, sunny weather, but at night heat is lost rapidly, leading to very cold conditions often with severe frosts, as there is no

contd

Physical Environments – Weather: High pressure systems (anticyclones)

blanket of clouds to trap heat. High pressure in winter is associated with calm, clear, sunny weather but intensely cold frosty nights. Often fog or even freezing fog forms early in the morning as the condensation level is close to the ground, but as the Sun rises this usually burns off quickly. If an anticyclone persists in winter, it can lead to weeks of clear, sunny days but bitterly cold frosty nights. The Sun is not high enough in the sky to give significant warmth and so temperatures remain low even during the day.

High pressure in summer (Whiten Head, Sutherland).

High pressure in winter.

ONLINE

Learn more about high and low pressure systems online by following the link in the BrightRED Digital Zone.

ONLINE TEST

Take the test on high pressure systems online at www.brightredbooks.net/N5Geography

THINGS TO DO AND THINK ABOUT

1. Watch the clip about the problems a summer anticyclone and heat wave could cause at www.brightredbooks.net/N5Geography

2. Copy and complete this table by adding two or three further examples to each section.

	Advantages	Disadvantages
Winter anticyclone	• clear sunny weather • • •	• very cold frosty nights • • •
Summer anticyclone	• nice weather for a visit to the beach • • •	• summer heatwave could cause drought • • •

PHYSICAL ENVIRONMENTS

LANDSCAPES: UPLAND GLACIATION

ICE AGE

During the last **Ice Age**, 11,000 years ago, nearly all of Britain was covered by **ice sheets**. Only the far south of England remained ice free. And that was only the most recent ice age - there have been several! So the landscape of the British Isles that we know today has been carved by a series of huge ice sheets and **glaciers**. The evidence is all around, from wide **U-shaped valleys**, such as the Great Glen between Fort William and Inverness, and the deep northern **corries** in the Cairngorms; to jagged **pyramidal peaks** and **arêtes** such as those found in the Cuillin Mountains on Skye.

VIDEO LINK

Watch the animation to see how glacial erosion can change the face of a landscape over millions of years at www.brightredbooks.net/N5Geography

GLACIERS

A **glacier** is like a huge, deep and very slow moving river of ice, but how is a glacier created? In winter snow builds up on mountain slopes and in summer, if it is not warm enough, some snow will remain in shaded hollows. With successive snowfalls over the following years, the snow builds up and the bottom layers have the air squeezed out of them and turn to a substance known as **névé** or **firn**. Eventually this will turn into ice, and with further snowfalls the ice will begin to flow downhill due to gravity. A glacier is born.

DON'T FORGET

Remember to explain these processes of glacial erosion if you are describing the formation of glacial features such as corries, U-shaped valleys, arêtes, etc.

GLACIAL EROSION

Where the ice builds up in mountain hollows there are three different processes at work. First, as water which has found its way into cracks and crevices in the rock starts to freeze, it expands by about 9% and helps to force the rock apart. When this happens repeatedly, the rock eventually shatters. This is known as **frost shattering** (and also **freeze-thaw action**). Second, as some of these fragments of rock are frozen into the base of the moving ice, they scrape and grind away at the land over which the ice moves, causing it to be eroded. This is called **abrasion**.

Third, as the ice moves it may freeze onto rock previously loosened by frost shattering and pull it from the ground as it continues slowly downhill. This is called **ice plucking**. Frost shattering, abrasion and ice plucking are the main ways in which ice erodes the landscape to create the spectacular features associated with glaciation.

DON'T FORGET

Corrie is a Scottish word. Corries are known as cirques in England and as cwms in Wales.

CORRIES

A **corrie** is formed high on the slopes of mountains where snow and ice have built up over many years. Usually this is in shaded, often north-facing hollows. As the snow and ice layers become deeper, the processes of glacial **erosion** increasingly affect the rock at the base of the hollow. Abrasion and plucking deepen the hollow as the ice moves over it and frost shattering loosens rock fragments, which are incorporated into the moving ice and lead to further abrasion. Slowly the hollow is deepened into a bowl shape as the glacier grows and a tongue of ice flows downhill.

When the ice has melted the bowl shape that has been carved into the mountainside is known as a corrie. Often there is a **rock lip** at the front of the corrie where hard resistant rock was scraped by fragments of ice in the base of the glacier. These scratches are called **striations**. The saucer shape at the base of the corrie often allows a small lake or **corrie lochan** to form there. At the back of the corrie there are very steep slopes or crags and cliffs where the

1 Snow accumulates on north-facing hollow on mountain

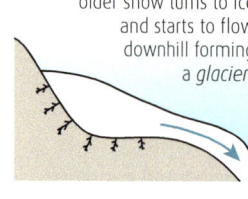
2 Snow continues to accumulate; older snow turns to ice and starts to flow downhill forming a *glacier*

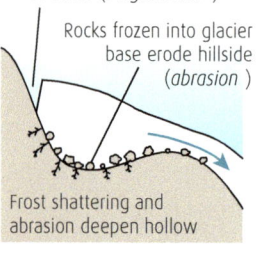
3 Gap forms at back wall of corrie (*bergschrund*)
Rocks frozen into glacier base erode hillside (*abrasion*)
Frost shattering and abrasion deepen hollow

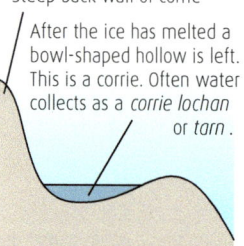
4 Steep back wall of corrie
After the ice has melted a bowl-shaped hollow is left. This is a corrie. Often water collects as a *corrie lochan* or *tarn*.

Diagram 1.18 A corrie.

contd

Physical Environments – Landscapes: Upland glaciation

ice eroded deeply into the land. **Scree** is found here due to frost shattering that continues to break up the rock every time the temperature falls below freezing point.

ARÊTES AND PYRAMIDAL PEAKS

Several corries often form on the slopes of large mountains. As the ridges of land between the corries are eaten into by the ice they become narrower, while frost shattering acts to make the ridges more jagged. After the ice melts, a mountain may have a number of corries separated by sharp rocky ridges. These are called **arêtes**. Where three or more arêtes lead up to the mountain top, they can create a very pointed summit, called a **pyramidal peak**. Scotland has many examples of pyramidal peaks, such as Ben Lui.

Ben Lui, Argyll: a pyramidal peak with arêtes.

U-SHAPED VALLEYS

As glaciers flow downhill due to gravity they follow the course of old river valleys. Glaciers flowing from many corries will join up to create a large valley glacier, which is immensely powerful. Contained within the glacier are millions of fragments of rock that the ice has picked up on its journey. This leads to very powerful abrasion on the bottom and sides of the valley, while plucking also helps the ice to carve downwards, widening and deepening it. If the temperature is warm enough, the end of the glacier melts faster than the ice is flowing and the **snout** of the glacier retreats back up the valley, revealing the newly enlarged glacial U-shaped valley with very steep sides and a wide flat bottom.

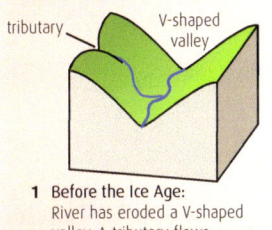 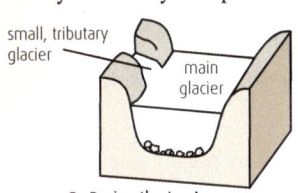 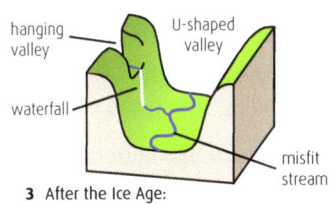

1 Before the Ice Age: River has eroded a V-shaped valley. A tributary flows down from the left.

2 During the Ice Age: A glacier flows down the valley, widening and deepening it due to processes of glacial erosion.

3 After the Ice Age: A steep-sided U-shaped valley is left with a wide flat bottom. The tributary now flows from a hanging valley high up on the left.

Diagram 1.19 A U-shaped valley.

Sometimes, where erosion has over-deepened the valley, a long narrow lake called a **ribbon loch** may form. Flowing along the floor of the **U-shaped valley** the river is known as a misfit stream, while high on the sides of the valley steep crags or **truncated spurs** may have formed, where the glacier has sliced off more gradual slopes that existed before. Sometimes a smaller **tributary** glacier that was less powerful than the main glacier, and so didn't erode as deeply, causes a smaller U-shaped valley to be left high up on the side of the main valley. This is called a **hanging valley** and the stream flowing from it often descends into the main valley by means of a **waterfall**.

Strath Dionard, Sutherland: a U-shaped valley.

ONLINE TEST

To test your knowledge of upland glaciation, visit www.brightredbooks.net/N5Geography

VIDEO LINK

Check out the clip about upland glaciation at Loch Lomond: www.brightredbooks.net/N5Geography

DON'T FORGET

The Scottish Highlands, with ranges such as the Cairngorms, have some excellent examples of glaciated scenery. It is useful to know at least one named example of each landscape feature you learn about.

ONLINE

Explore this topic further by following the 'Glacial landforms' link in the BrightRED Digital Zone.

THINGS TO DO AND THINK ABOUT

1. Use Google Maps or look at a road atlas of Scotland to find the mountainous areas above 750 metres. Search for the named examples of glacial features that have been mentioned on these pages.

2. In a world atlas find these mountainous areas, which are all examples of glaciated areas: the Lake District, England; Snowdonia, Wales; the Alps, Switzerland; the Rocky Mountains, USA and Canada.

PHYSICAL ENVIRONMENTS
LANDSCAPES: COASTS 1

Britain's coastline, including the main islands, is about 19,491 miles long and nowhere is more than 70 miles from the sea. Being an island nation, our landscape has been shaped by the sea for millions of years. Two principal types of coastline can be identified: coastlines of erosion and coastlines of deposition.

ONLINE
Learn more by following the 'Erosion' link in the BrightRED Digital Zone.

Diagram 1.20 Fetch and prevailing wind direction.

VIDEO LINK
Check out the 'Coastal erosion' clip at www.brightredbooks.net/N5Geography

COASTLINES OF EROSION

The amount of erosion that happens along the coast depends on a number of factors, such as the rock type, weather conditions and the fetch. The **fetch** is the distance over which waves build up. The bigger the fetch the larger the waves. The most common wind direction affecting the British Isles, known as the **prevailing wind**, is south-westerly. This allows waves to build up over the whole width of the Atlantic Ocean and they are therefore large and powerful when they hit the western coastline of the British Isles, causing extensive erosion.

There are several ways in which the sea erodes the land:

- **Wave pounding** is where the weight of water hitting the cliff face repeatedly erodes it. Large breaking waves can exert pressure of 30 tonnes per square metre.
- **Hydraulic action** happens when air is trapped by waves in cracks and crevices on the cliff face, causing it to be compressed and then released with explosive force.
- **Corrasion** is when stones and fragments of rock are flung against cliffs by the breaking waves and erode them.
- **Corrosion** is where rock is dissolved by carbonic acid found in sea water.
- **Attrition** is where stones and boulders swirled around by the waves are gradually broken up and made smaller. The end product of this process is sand.

COASTAL EROSION LANDFORMS

When waves attack a coast, they do so most intensely at the base of cliffs. Repeated erosion here, which is greatest at high tide and during storms, creates a **wave-cut notch**. Over time the notch gets bigger, undermining the cliff and making it unstable. The cliff collapses and the process begins again, causing the coast to move back and creating a **wave-cut platform**, which is what is left at the base of the former cliffs.

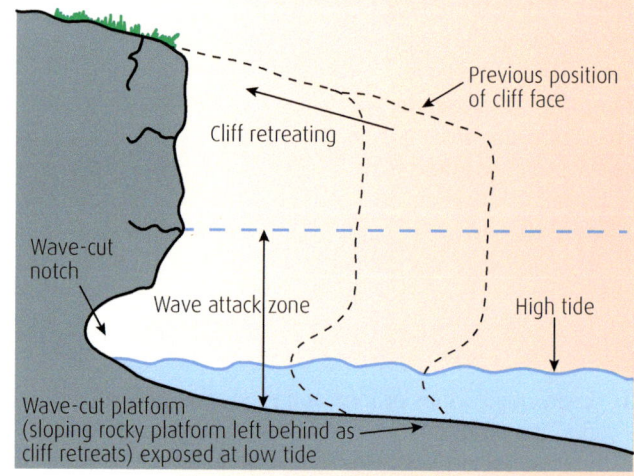

Diagram 1.21 Retreating cliffs.

contd

Physical Environments – Landscapes: Coasts 1

If the coastline consists of alternate bands of hard and soft rock, **differential erosion** will take place. This is where the sea erodes the areas of soft rock more quickly because they are less resistant. This can lead to the creation of **bays**, where there is soft rock, and **headlands** extending out beyond the bays, where there is harder rock.

On the headlands the processes of coastal erosion often produce the most spectacular coastal landforms. Erosion quickly acts on weaknesses in the cliff such as fault lines. Before long, because of the processes of erosion listed, these can be hollowed out to form **caves**. If the roof of the cave is eroded all the way to the surface, a **blowhole** is formed. In stormy weather, spray may be ejected form the blowhole just like a spouting whale. Occasionally, caves can extend all the way through to the other side of the **headland** to form a **sea arch** (one of the best known examples being the Green Bridge of Wales). With further erosion the arch may eventually collapse to leave a **pillar** of rock known as a **sea stack** (the largest of these in Britain is the Old Man of Hoy, at 137 metres high). Finally the sea stack too will succumb to erosion and collapse into the sea, leaving a **stump**.

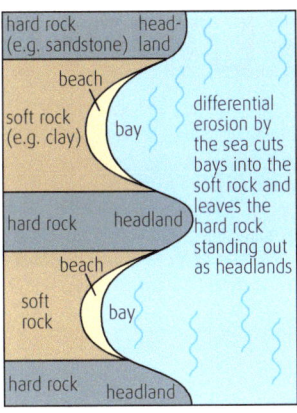

Diagram 1.22 Headlands and bays.

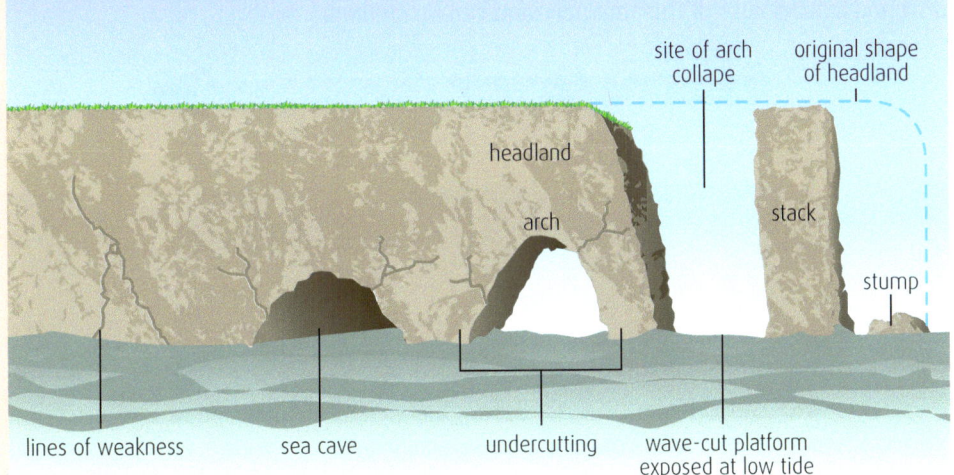

Diagram 1.23 Coastal erosion features.

Coastline of erosion at Yesnaby, Orkney.

Sea caves at Yesnaby.

DON'T FORGET

Coastal erosion is most rapid at high tide, when there are strong winds blowing on shore. Other factors affecting the rate of erosion include the rock type and the width of the beach. Wide beaches absorb wave energy.

VIDEO LINK

Check out the clip about eroding coastlines at www.brightredpublishing.co.uk/N5Geography

VIDEO LINK

Watch the animation of cliffs retreating at www.brightredbooks.net/N5Geography

THINGS TO DO AND THINK ABOUT

1. Use a search engine to find pictures of some of the best-known coastal erosion landscapes in Britain. Look for **the Green Bridge of Wales, Old Harry Rocks, Elephant Rock near Montrose and the Old Man of Hoy**.

2. On an outline map of the British Isles, plot the locations of the places above and label them appropriately as **sea stack** or **sea arch**.

3. Practise drawing diagrams of coastal erosion features. Add explanatory labels to show how each feature is formed.

PHYSICAL ENVIRONMENTS
LANDSCAPES: COASTS 2

COASTLINES OF DEPOSITION

Although the sea is a powerful agent of erosion, it can also deposit material and create new land. The main process which causes this to happen is the process of **longshore drift**. When waves break diagonally up a beach according to the direction of the prevailing wind, they push beach sediment with them. This breaking wave is called the **swash**. When gravity causes the water to flow back down the beach vertically, it also drags beach sediment back down with it. This is called the **backwash**. The effect of this happening repeatedly is that the sediment (often sand) moves along the beach in the same direction as the prevailing wind is blowing. This process is known as longshore drift and it is because of this that new land can be created.

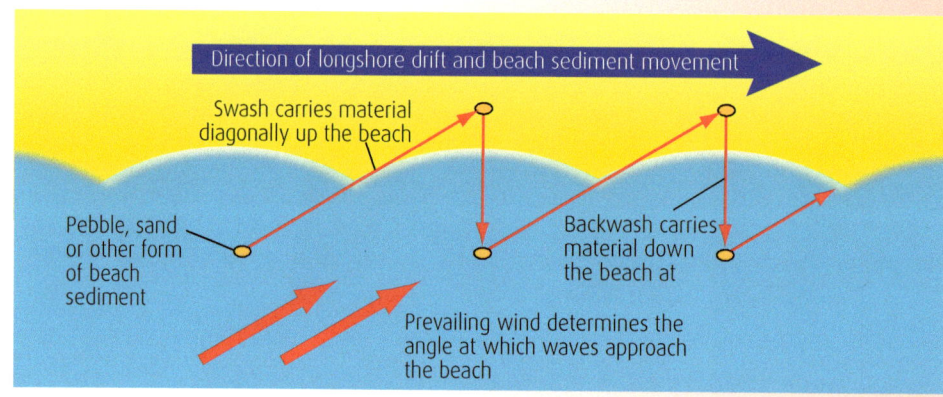

Diagram 1.24 Longshore drift.

VIDEO LINK

Check out the 'Longshore drift' clip at www.brightredbooks.net/N5Geography

DON'T FORGET

The prevailing wind is the direction from which the wind blows most frequently.

ONLINE

Learn more by following the 'Coasts of erosion' link at www.brightredbooks.net/N5Geography

COASTAL DEPOSITION LANDFORMS

If there is a bend in the coastline, longshore drift may continue to follow the direction of the prevailing wind, causing sand to be deposited out to sea and creating a new peninsula of land known as a **sand spit**. Sometimes, when the wind is blowing from the second most common direction, the direction of longshore drift changes, causing the sand spit to grow temporarily in a different direction. This creates a **hooked** or **recurved spit**. In the slack water behind the hooked spit a **salt marsh** is often formed where shallow water is colonised by marsh vegetation, which traps more sand, eventually creating new land.

A tombolo at St Ninian's Isle, Shetland.

contd

Physical Environments – Landscapes: Coasts 2

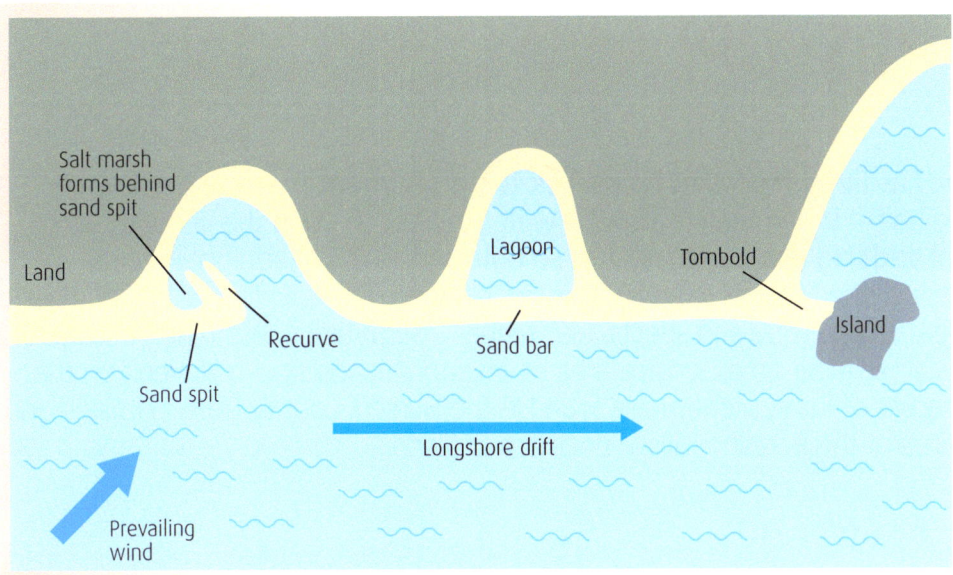

Diagram 1.25 Diagram of coastal deposition features.

If a sand spit grows across a **bay** and joins with the opposite shore, it creates a feature known as a **sand bar**, sealing off the bay and forming a **lagoon**. The lagoon, like the salt marsh, may eventually fill with marsh vegetation, trapping more sand and creating new land. A further landform caused by **deposition** is when a sand spit joins an offshore island, creating a feature known as a **tombolo**. A good example of a tombolo is found at St Ninian's Isle in Shetland.

In Britain it is unusual for coastal deposition landforms to grow far out to sea because of deep water, powerful offshore currents and strong winds. However, Spurn Head spit at the mouth of the Humber **estuary** is a recurved spit that has grown to over 6 miles in length.

Sand spit at Spurn Head, Humber estuary.

THINGS TO DO AND THINK ABOUT

1. Use **Google Earth** to find some of Britain's best-known coastal deposition features. Look for **Chesil Beach Weymouth**, **Slapton Ley**, **St Ninian's Isle** and **Spurn Head**.
2. On an outline map of the British Isles, plot the locations of the places above and label them as lagoon and sand bar, sand spit, tombolo. There are two examples of one feature.

ONLINE TEST

How much do you know about coastlines of deposition? Test yourself at www.brightredbooks.net/N5Geography

PHYSICAL ENVIRONMENTS

LANDSCAPES: UPLAND LIMESTONE

Carboniferous limestone has three important properties. It is a **sedimentary** rock, laid down in layers on an ancient sea bed. It is permeable, allowing water to pass through the vertical joints and along the horizontal **bedding planes**. Finally, it is **soluble** and, in mildly acidic rainwater, carboniferous limestone will slowly dissolve away. This combination of properties results in areas of carboniferous limestone having unique and spectacular scenery. Sometimes, this type of scenery is referred to as **karst**, named after an extensive area of limestone landscape in Slovenia. In the British Isles the largest areas of karst scenery are found in the Burren, County Clare in Ireland and in the Yorkshire Dales area of the Pennine Hills in northern England.

CARBONATION

One of the unique properties of **carboniferous limestone** is its solubility. As it falls through the atmosphere, rain absorbs carbon dioxide, making a weak carbonic acid. Limestone is made of calcium carbonate and is strongly alkaline, so it reacts with the acidic rainwater, creating calcium hydrogen carbonate, which is washed away in solution by running water. The chemical formula for this process of **carbonation** is:

$$CaCO_3 + H_2CO_3 = Ca(HCO_3)_2$$

calcium carbonate + carbonic acid = calcium bicarbonate
(limestone) (rain water) (soluble limestone)

This process is vital to understanding the unique surface and underground landscapes of carboniferous limestone areas.

SURFACE FEATURES

VIDEO LINK

Watch the 'Surface features' clip for more at www.brightredbooks.net/N5Geography

In the Yorkshire Dales carboniferous limestone outcrops are found amongst areas of **millstone grit**, which is an **impermeable** rock. Water flows in streams and rivers above the impermeable millstone grit, but once these rivers reach the **permeable** carboniferous limestone they are liable to disappear underground down an enlarged vertical **joint**, known as a **swallow hole**. Once underground, flowing water creates an extensive network of subterranean streams and cave systems. Sometimes streams will reappear from underground at a **resurgence** to flow on the surface once again where the limestone meets another layer of impermeable rock. This pattern of disappearing and reappearing streams is a common feature of **karst scenery** and is known as **intermittent drainage**.

Glaciokarst is the term used to describe the situation where large areas of carboniferous limestone have been exposed by the scouring action of glaciers and ice sheets, which removed layers of rock and soil from above them. Once open to the elements, these areas of exposed limestone were subject to the effects of weathering. Vertical joints in the limestone were affected by carbonation and have become enlarged to form miniature ravines known as **grykes** between the remaining blocks of limestone, known as **clints**. This combination of grykes and clints forms a feature known as a **limestone pavement**. The grykes can be easily a metre or more in depth and provide shelter for calcium tolerant plants, which grow in between the clints.

ONLINE

Explore this more by following the 'Surface features' link at www.brightredbooks.net/N5Geography

Water is rarely found on limestone pavements because of its permeability. Any streams flowing onto the surface of limestone pavements disappear down swallow holes or **water sinks**. A famous example of a swallow hole is found at Gaping Gill on the slopes of Ingleborough in the Yorkshire Dales, where the disappearing stream creates an underground waterfall some 98 metres high. Also found on areas of limestone pavement are depressions that may be several metres deep and are full of **boulder clay** or other debris. These are known as **shake holes** and can be hazardous for walkers.

contd

Glaciers are also responsible for creating **limestone scars**, which are large areas of exposed limestone often found on the sides of glaciated valleys in upland limestone areas. **Frost shattering** results in the accumulation of **scree slopes** at the base of these scars.

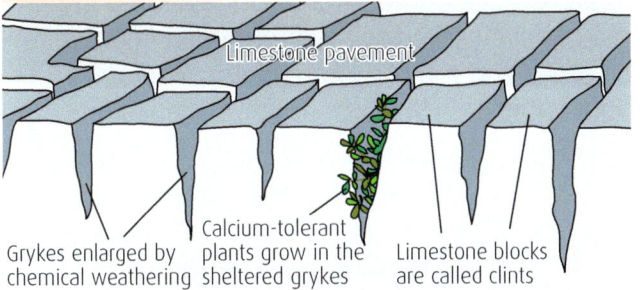

Diagram 1.26 Limestone pavement.

Limestone pavement at The Burren, Ireland

UNDERGROUND FEATURES

Flowing water underground creates a network of caves and subterranean streams. As water disappears down a swallow hole, it erodes and dissolves passageways for itself along existing bedding planes and joints. Eventually these are widened out into underground **caves** and **caverns**. Erosion by flowing water as well as solution of the calcium carbonate are both involved in the creation of caves. **Phreatic caves** form beneath the water table where water is constantly in contact with all the surfaces of the cave so that they are evenly eroded and often tubular in shape. **Vadose caves** are above the water table and less regular as water is not in contact with all the surfaces. Some cave systems are vast and some have been opened up to visitors as show caves, such as White Scar Caves in Yorkshire, which claims to be the longest show cave in Britain.

Dripstone features are found in vadose caves. This is when water dripping slowly from the roof of a cave evaporates, leaving behind a small deposit of **calcite**. This comes from the limestone that has been dissolved in the water higher up in the limestone. Over time, as this crystallisation continues, icicle-shaped deposits called **stalactites** form, hanging from the cave roof. **Straw stalactites** can also form. These are particularly thin, fragile and hollow. Stalactites form very slowly at no more than about 3 **millimetres** per year. Where drips land on the cave floor, evaporation also takes place and mounds of calcite form sticking up into the cave. These are called **stalagmites** and are more rounded and stumpier than stalactites. When a stalactite and stalagmite join together a **pillar** or **column** is formed.

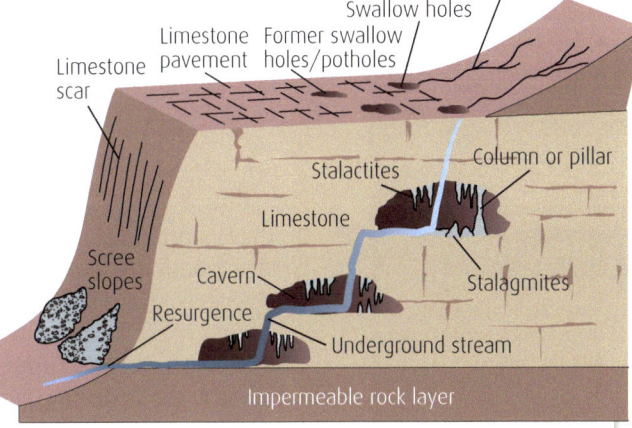

Diagram 1.27 Limestone (karst) landscape.

Subterranean streams continue to flow down through the joints and bedding planes in the limestone until they reach a layer of impermeable rock. They will flow along this until sometimes they reappear on the surface at a **resurgence** where the limestone meets the impermeable rock.

ONLINE
Explore underground features more at www.brightredbooks.net/N5Geography

DON'T FORGET
Karst scenery is a term often used to describe the scenery in (carboniferous) upland limestone areas.

ONLINE TEST
How much do you know about upland limestone? Test yourself at www.brightredbooks.net/N5Geography

THINGS TO DO AND THINK ABOUT

1. Write down all the limestone landscape features that appear in bold type on these pages in two columns, one for surface features and one for underground features.
2. Watch the BBC Horizon documentary 'The Secret Life of Caves' at www.brightredbooks.net/N5Geography

PHYSICAL ENVIRONMENTS
LANDSCAPES: RIVERS

The impact of rivers on our landscape is often overlooked. Although Britain's longest river, the Severn, is only 220 miles from source to mouth, rivers have carved some of our most impressive scenery. From deep V-shaped valleys to wide flood plains and tumbling waterfalls, rivers are responsible for much of the way our landscape looks today.

RIVER EROSION

There are four main ways in which rivers erode the landscape.

1. **Corrasion:** the river's **load** of stones and pebbles scrapes away at the river's bed and banks, causing them to be worn away.
2. **Hydraulic action:** the power and pressure of the flowing water itself can wear away the river banks.
3. **Corrosion:** weak acids in river water allow chemical reactions to take place on the river's bed and banks, causing them to be eroded.
4. **Attrition:** as a river flows from the uplands to its mouth, the pebbles and stones it carries continuously collide with each other, becoming smaller. By the time a long river reaches its **mouth**, most of its load has been reduced to small pieces of gravel, sand or mud due to attrition.

Diagram 1.28 River: upper, middle and lower course.

RIVER LANDSCAPES

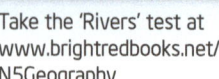

ONLINE TEST

Take the 'Rivers' test at www.brightredbooks.net/N5Geography

Rivers start in upland areas where there is high rainfall. The point furthest from the river mouth is known as the **source**. Water starts to flow downhill due to gravity and what started as a small stream is quickly joined by others, called **tributaries**, to become a larger river. The point at which two rivers join together is called a **confluence**. When a river is in its **upper course**, it flows quickly down relatively steep gradients and has the power to erode deeply into its bed. A **V-shaped valley** is formed when the river erodes downwards by corrasion and hydraulic action, while **weathering** on the valley sides loosens material that falls into the river and is carried away by the water.

V-shaped valley.

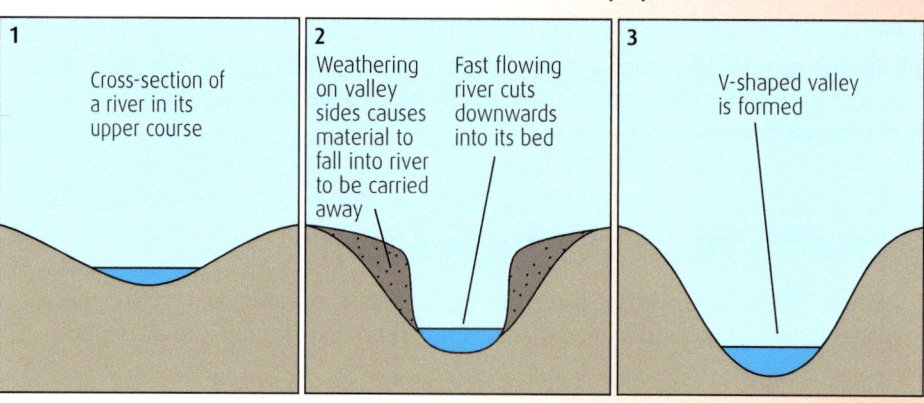

Diagram 1.29 Formation of a V-shaped valley.

contd

28

Physical Environments – Landscapes: Rivers

A common feature in a river's upper course is a **waterfall**. This is caused by the river flowing over an especially hard band of rock that is resistant to erosion. Where it flows back onto a softer rock band, it is able to erode downwards more quickly and so a step is formed. Over time, this step becomes bigger and the river flows over a significant drop, creating a waterfall. Through the processes of corrasion and hydraulic action, the falling water in the **plunge pool** erodes the soft rock, undercutting the hard rock band. Without support, the hard rock band eventually collapses and the position of the waterfall moves backwards. As this happens repeatedly over time, a steep-sided narrow valley, known as a **gorge**, is formed.

As the river enters its **middle course**, the gradient is less steep and the river has less erosive power. It flows through a wider, more gentle valley and often has a **flood plain** on either side. Often the river winds from side to side, creating river bends known as **meanders**. In the **lower course** of a river there is less erosive power and as it has less energy, the river may deposit material on the bed and banks. Through a combination of erosion and deposition, a feature known as an **ox-bow lake** can be formed at a meander. The process by which this happens is as follows.

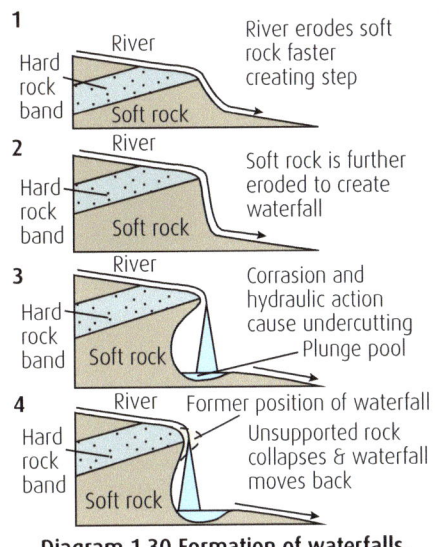
Diagram 1.30 Formation of waterfalls.

> **DON'T FORGET**
>
> Waterfalls resulting from a band of hard rock in a river valley are formed in a different way from waterfalls found at the end of a glaciated hanging valley.

> **VIDEO LINK**
>
> Learn more by watching the 'Levées' clip at www.brightredbooks.net/N5Geography

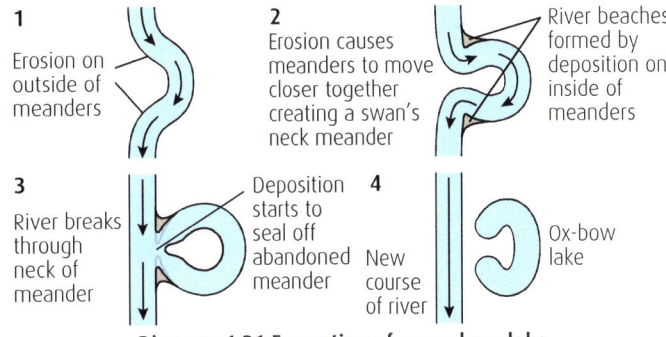
Diagram 1.31 Formation of an ox-bow lake.

1. The faster flowing current on the outside bends of the meander erodes the banks, while deposition happens where the water is slower flowing and less powerful on the inside bends of the meander.

2. Erosion on the outside bends continues until **a swan's neck meander** is formed.

3. Continual erosion on the outside bends eventually results in the river breaking through the neck of the meander and taking a new, shorter route.

4. Deposition on the inside bends of the meander eventually seals off the old meander, creating an ox-bow lake. Over time this lake may become overgrown with reeds and rushes and, in time, become new land.

The course of a river changes significantly over time. Ox-bow lakes can be formed and then disappear. Meanwhile, the river meanders over its flood plain, often eroding the edges and making it wider.

Another feature of a river in its lower course, before it finally reaches its mouth and flows into the sea, are **levées**. These are natural banks formed on either side of the river resulting from repeated flood action. If a river bursts its banks as a result of particularly heavy rain, the water which flows on to the flood plain initially slows down because of increased friction and deposits some of the material it is carrying (known as the river's load) next to the river. Over time these deposits can build up into mounds running parallel to, and on either side of, the river. These are levées. On some large rivers such as the Mississippi, levées can be several metres high.

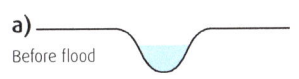
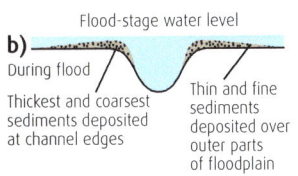
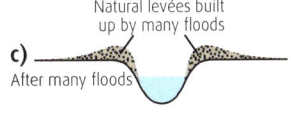

Diagram 1.32 Formation of levées.

> **ONLINE**
>
> Explore rivers further online at www.brightredbooks.net/N5Geography

 THINGS TO DO AND THINK ABOUT

Study the labelled diagrams of V-shaped valleys, waterfalls, ox-bow lakes and levées. Close the book and practise drawing them from memory.

PHYSICAL ENVIRONMENTS
LANDSCAPES: O.S. MAP SKILLS 1

COASTS

Britain has a long coastline that is one of the most varied in the world. Being able to recognise different coastal features on Ordnance Survey (O.S.) map extracts is a skill you are expected to have. The following diagrams highlight some of the different features that you should be able to recognise and identify. They include landscapes of coastal erosion as well as deposition. Some features, such as caves or sea arches, don't have a particular symbol and are often just indicated with the word 'cave' or the name of an arch, e.g. Durdle Door written on a blue background, indicating a feature of tourist interest.

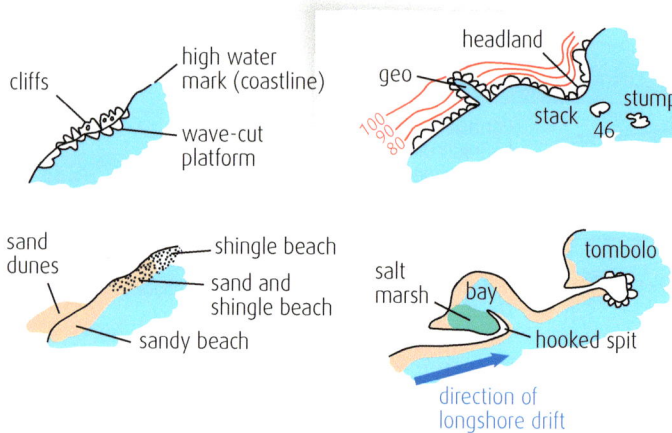

Diagram 1.33 Recognising coastal landforms on O.S. maps.

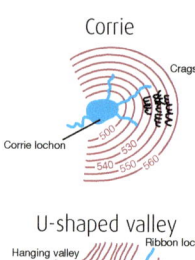

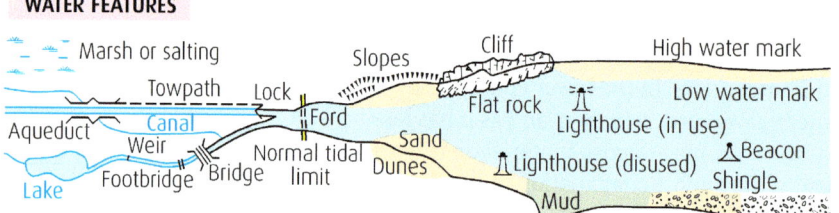

Diagram 1.34 Key to water features on O.S. maps.

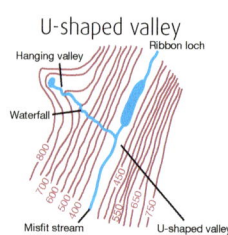

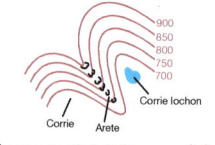

Diagram 1.36 Recognising glaciated landforms on O.S. maps.

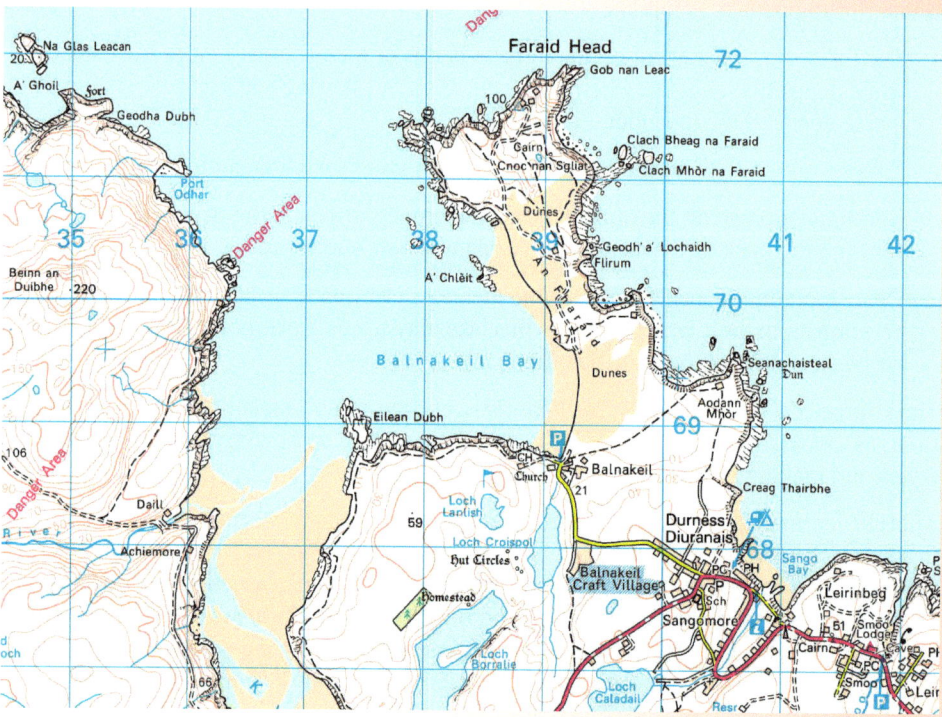

Diagram 1.35 1:50,000 map extract of Durness coast area.

UPLAND GLACIATED LANDSCAPES

Upland glaciated areas have many spectacular features that are clearly evident on O.S. maps. **Corries** are characterised by a distinctive horseshoe-shaped pattern of **contour** lines, enclosing a hollow with three very steep sides, sometimes with crags and cliffs. Often there is a lochan or tarn in the bottom of the corrie. A **U-shaped valley**

contd

appears as a huge valley with a misfit stream in the bottom, or perhaps even a **ribbon loch**, surrounded by a little flat land but with very steep slopes on either side. **Arêtes** appear as sharp ridges where the contours double back on themselves (as they show the start of the next corrie). Often the symbols for bare rock and **scree** appear, giving evidence of frost shattering.

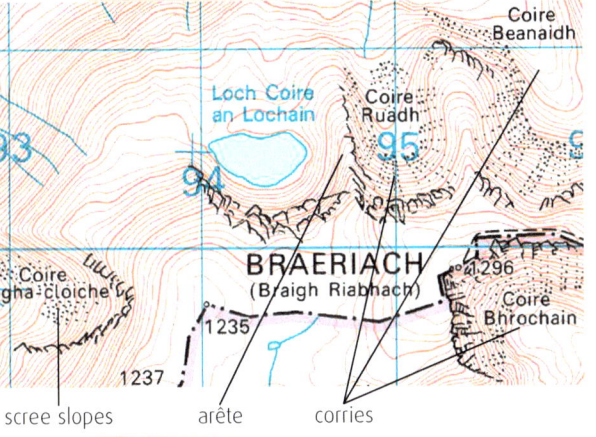

Diagram 1.37 1:50,000 map extract of Braeriach Cairngorms.

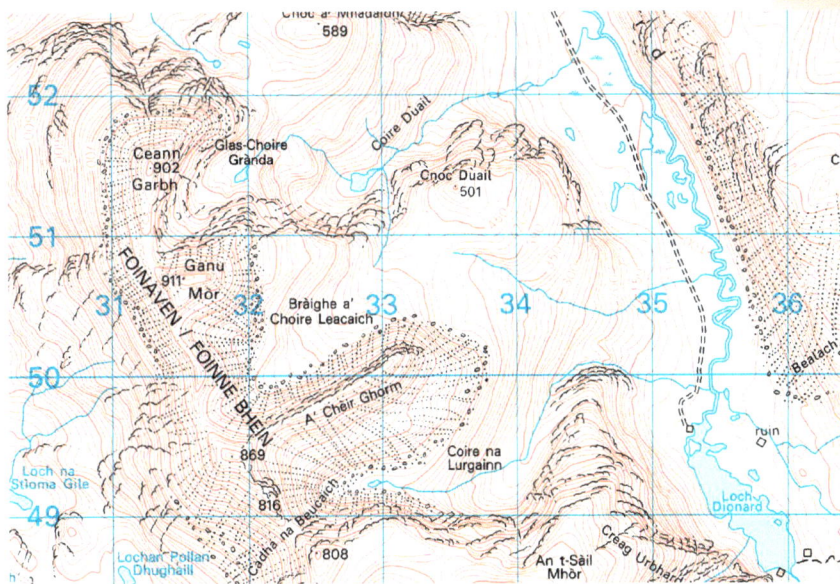

Diagram 1.38 1:50,000 map extract of Foinaven area.

 THINGS TO DO AND THINK ABOUT

1. Match the grid references in the table with the correct coastal features from the map extract of Durness (Diagram 1.35). Choose from:

   wave-cut platform   **sandy beach**   **sea stack**   **headland**   **cave**

Grid reference	Coastal feature
347720	
401694	
419672	
392695	
350717	

2. Study the map extract of Durness and write down a list showing **other** examples of coastal landforms. Write a six-figure grid reference next to each feature you identify. Landscape features you could look for are:

   sand dunes   **geos**   **cliffs**   **sea stacks**   **stumps**   **wave-cut platforms**

3. Match the grid references in the table with the correct glaciated features from the map extract of Foinaven (Diagram 1.38). Choose from:

   corrie   **U-shaped valley**   **arête**   **ribbon loch**   **hanging valley**

Grid reference	Glaciated feature
333516	
357490	
325498	
347520	
318514	

4. Study the map extract of Foinaven and write down a list showing other examples of glaciated features. Write a six-figure grid reference next to each feature you identify. Landscape features you could look for are:

   scree slopes   **misfit stream**   **corries**   **corrie lochans**

 DON'T FORGET

Being able to recognise landscape features on a map is a really useful skill, not just for the exam but also if you are navigating outdoors.

 ONLINE TEST

Take the 'O.S. map skills: Coasts' test at www.brightredbooks.net/N5Geography

 ONLINE

Check out the 'Map reading information' link for more at www.brightredbooks.net/N5Geography

 VIDEO LINK

Check out the 'Understanding map symbols' clip online at www.brightredbooks.net/N5Geography

PHYSICAL ENVIRONMENTS
LANDSCAPES: O.S. MAP SKILLS 2

VIDEO LINK

Check out the 'Learn about contour lines' clip online at www.brightredbooks.net/N5Geography

RIVERS

When describing a river on an O.S. map it is important to look at the *physical* (natural) features of the river and its valley. Ask yourself these questions about the river:

- Is the river in its **upper**, **middle** or **lower course**?
- What is the shape of the valley? Lots of contours suggest a **steep V-shaped valley**; few contours may suggest a **gently sloping valley** in the lower course.
- Is there land on either side without contours? If so, may be a **flood plain**.
- Can you identify river landforms (e.g. **meanders**, **ox-bow lakes**, **tributaries**, etc.) with six-figure grid references?
- What is the **name** of the river and the names of any tributaries?
- In which **direction** does the river flow?
- How deep/wide is the valley? Use the contour heights to work this out.

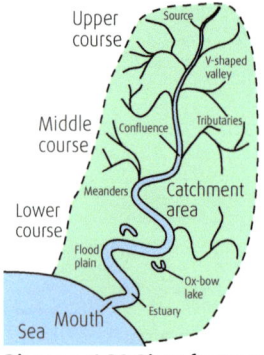
Diagram 1.39 River features.

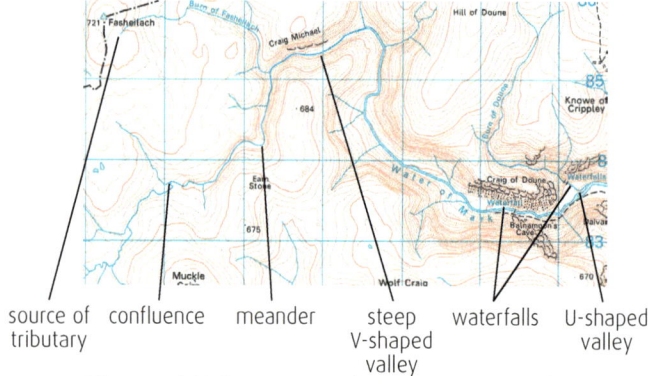
Diagram 1.40 Upper course river: 1:50,000 map of river.

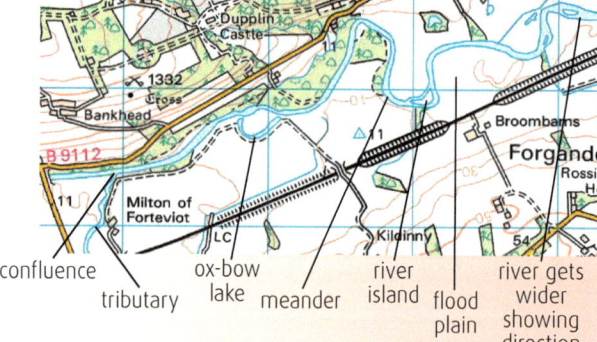
Diagram 1.41 Lower course river: 1:50,000 map of the River Earn.

In a river's upper course, look for features such as the **source**, steep V-shaped valleys, tributaries, confluences, **waterfalls** and **gorges**. Tightly packed contours will indicate that the river has a steep gradient and is likely to be in its upper course. Usually, there is no flat land in the valley bottom as the valley sides rise steeply from close to the river.

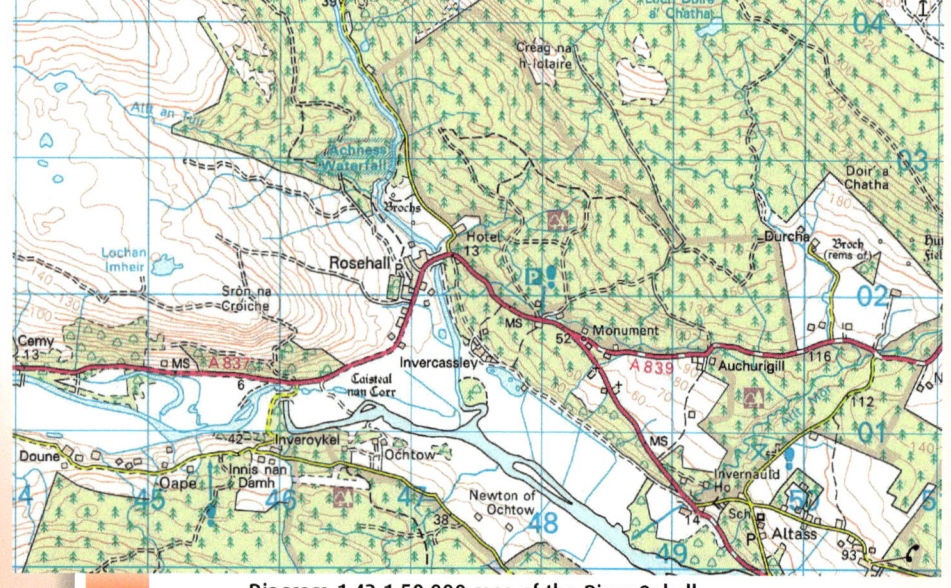
Diagram 1.42 1:50,000 map of the River Oykell.

In a river's lower course, the river channel is usually wider and the landscape is more gentle, with wide **flood plains** and broad **meanders**. An absence of contours on one, or both, sides of the river indicates the presence of a flood plain. **Ox-bow lakes**, meanders and **levées** are also more likely to be encountered in the lower course of a river. The **tidal limit** of a river is shown by the outline of the river turning from blue to black. This marks the furthest point up to which sea water reaches at high tide. A section of river outlined in black shows that there is sea water and therefore that section is part of the river **estuary**.

Physical Environments – Landscapes: O.S. map skills 2

LIMESTONE

An area of carboniferous limestone will be evident on an O.S. map due to a number of unique features. Streams might disappear underground and then reappear on the surface further downhill. This indicates **intermittent drainage**, with streams flowing down **sink** or **swallow holes** and reappearing at **resurgences**. There may be **potholes** or **shakeholes** marked, and caves may also be named on the map. Areas of **limestone pavement** are shown by the rock symbol in areas where there are few contours and there may also be **scree slopes** and **limestone scars** on valley sides close by. **Dry valleys** appear on the map as normal valleys but without a stream in the bottom.

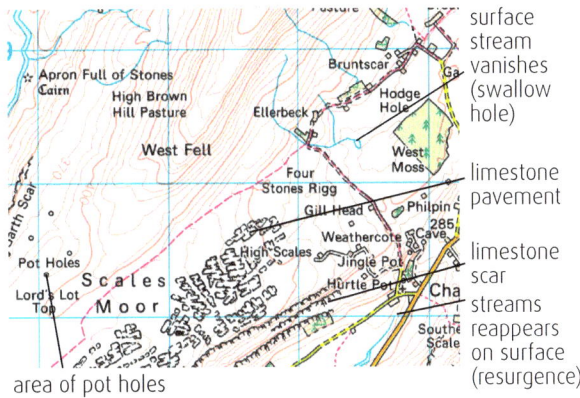

Diagram 1.43 Carboniferous limestone area on 1:50,000 map.

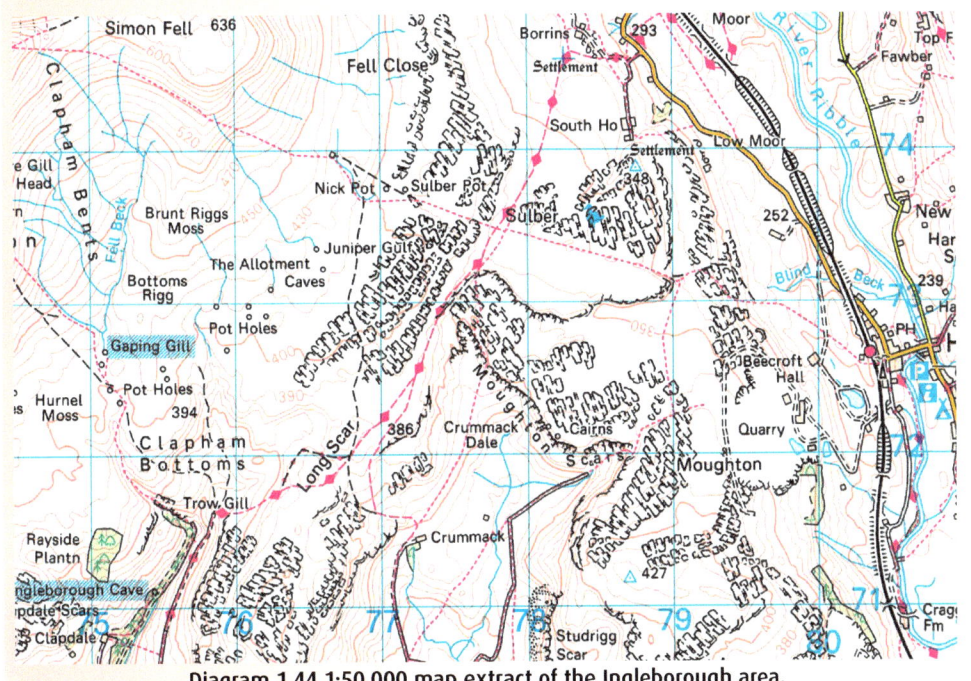

Diagram 1.44 1:50,000 map extract of the Ingleborough area.

ONLINE TEST

Take the 'O.S. map skills: Rivers and limestone' test at www.brightredbooks.net/N5Geography

DON'T FORGET

Only surface features are visible on a map of a limestone area. They suggest that there will be a vast network of underground features too.

THINGS TO DO AND THINK ABOUT

1. Match the grid references shown below with the correct river features on Diagram 1.42.

   ox-bow lake   tidal limit of river   confluence   waterfall   meander   floodplain

   Grid references:
   468029 _____   455011 _____
   475010 _____   461013 _____
   476012 _____   451014 _____

2. Describe the course of the River Oykell as it flows from grid square 4401 to 4800. Mention the valley as well as any river features you can identify.

3. Match the grid references shown below with the correct limestone features on Diagram 1.44.

   swallow hole   limestone pavement   limestone scar   resurgence   cave

   Grid references:
   754711 _____   782721 _____
   751727 _____   777719 _____
   770730 _____

4. Write a paragraph giving evidence to show that Diagram 1.44 shows an area of carboniferous limestone scenery. Try to include examples other than those in question 3.

PHYSICAL ENVIRONMENTS

LAND USES: FARMING AND FORESTRY

Agriculture uses up to 73% of the land in the UK, with 52% being used for rough grazing or pasture land and 21% being used for **arable farming**.

We depend on the farming industry, as some 60% of all the food we eat in the UK is produced here. Farming, in all its different forms, is therefore a vitally important land use. There are many different human and physical factors that affect the type of farming that might take place in a particular location.

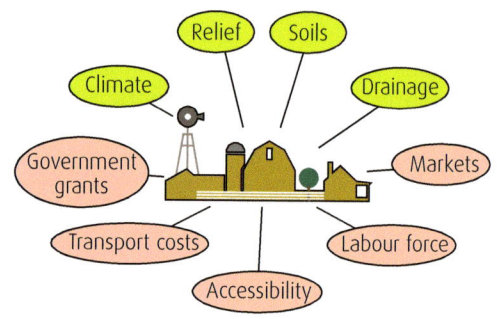

Diagram 1.45 Physical and human factors affecting farming.

FARMING IN GLACIATED UPLANDS

Blackface sheep can survive the harsh conditions on Scotland's mountains.

Native breeds such as Highland cattle are farmed on lower slopes and valley bottoms.

Farming in the bottom of the U-shaped valley at Glen Clova.

The landscape in glaciated upland areas is not particularly favourable for agriculture because the **relief** and altitude of the land are problematic. In the highest parts of Scotland's glaciated landscapes, such as the Cairngorm Mountains at over 1000 metres above sea level, the climate is too severe even for sheep farming. Other land uses, such as nature conservation, and leisure activities, such as skiing and hill walking, dominate. However, lower down and away from the main recreation and conservation zones, sheep farming is possible as the hardiest breeds are able to live out on the hillsides. In the Angus Glens, such as Glen Clova, hardy **black-face sheep** survive out on the hillsides from May until September where enough grass grows among the rocks and heather for them to feed off. In the winter months, they are kept lower down or on better pastures further down the valleys. In the valley bottoms, there are better **alluvial** soils mixed with boulder clay, and cattle farming is possible here. Pedigree cattle and sheep are bred here, although any crops that are grown are mainly used as **fodder** for the livestock. **Organic farming** is becoming more popular as demand for organic products increases, and many farmers have taken advantage of grants for participating in **environmental stewardship** schemes. Farms in glaciated areas may also be well placed to benefit from tourist-related forms of diversification, for example, running a farmhouse bed and breakfast or renting out holiday cottages. Being situated in scenic glaciated mountain areas can therefore be an advantage.

FARMING IN RIVER VALLEYS

In a similar way to that of glaciated upland areas, farming in river valley landscapes differs according to the relief and altitude of the land. Higher areas in the upper course of a river valley might be suitable for sheep farming, with a minimal amount of arable farming taking place on the more sheltered land and better alluvial soils of the valley bottoms. However, as the river valley enters its middle and lower courses, the nature of farming changes as soils, climate and relief all become more favourable. Flood plains are often used for arable farming because of their rich alluvial soils and flat topography, which makes it easier for machinery to operate. However, on low flat land, drainage, along with flooding, can be a problem and so arable farming is often best suited to the more gentle slopes of the valley sides in the lower course of a river.

ONLINE

Learn more about forestry and farming in glaciated areas online at www.brightredbooks.net/N5Geography

FORESTRY IN GLACIATED AREAS

Over 17% of Scotland's land is covered in forest. Much of it is **commercial plantations** (trees planted to produce timber in the future) owned by private companies or by the Forestry Commission. The most common types of trees are coniferous species such

contd

Physical Environments – Land uses: Farming and forestry

as fir, pine or spruce. Scotland's native coniferous tree is the **Caledonian pine** but there are only small areas of this natural forest left such as at Rothiemurchus in the Cairngorm Mountains. The most common type of commercially planted tree is the **Sitka spruce**. The UK **imports** much of its timber requirements from Canada, Norway and Sweden but the aim is to make Britain more self-sufficient in timber by planting more trees. This could increase the amount of land in Scotland covered in forest to 25%. Planting more trees is called **afforestation** and Diagram 1.46 shows some of the advantages and disadvantages of this.

Glaciated areas can be good for planting trees because there are fewer competing land uses than in lowland areas. Coniferous trees can also survive the lower temperatures, harsher climate and poor soils of upland areas, although only the lower slopes are suitable for planting as the higher land is too windswept and hostile. However, some groups of people are opposed to afforestation because they believe it blocks access to land for other uses and because uniform stands of coniferous trees don't look natural.

Diagram 1.46 Problems and benefits of forestry.

There are lots of possible recreational uses for forests and these are seen as increasingly important in the 21st century. Car parks, picnic areas, forest trails and information centres are provided for visitors. In many areas, such as Glenmore Forest Park in the Cairngorms, there are camping and caravan sites, as well as bird observatories to view ospreys at Boat of Garten in the nearby Abernethy Forest.

Road access is an important factor in commercial forestry.

There is also provision for many specialist activities such as mountain biking, with a variety of graded courses at Glentress Forest in the Scottish Borders. Other activities include orienteering, rallying, fishing and pony trekking.

FORESTRY IN RIVER VALLEYS

The slopes of river valleys in particular may be suited to forestry because there are only a limited number of alternative land uses. In the valleys of rivers in their upper and middle courses, the lower slopes may be used for commercial forestry plantations. Here the cooler temperatures and higher rainfall make forestry a suitable land-use option. The angle of slope is important so that machinery is still able to work (for example, to harvest the timber) and good road access is needed to be able to transport timber out of the area. Often forestry companies will construct substantial road networks of their own for this purpose. Other land uses, such as farming and **settlement** on the valley floors and floodplains, mean that forestry is less common here. The flatter land and more favourable climate in the lower course of a river are also likely to result in forestry being only a minor land use as others predominate.

 VIDEO LINK

Find out more about the diversification of forestry in the Tweed Valley Forest Park by watching the clip at www.brightredbooks.net/N5Geography

 VIDEO LINK

Learn more by watching the clip about land use at Loch Lomond at www.brightredbooks.net/N5Geography

 DON'T FORGET

Although glaciated uplands and river valleys have physical problems related to relief, climate and soil quality, they are far from useless. People have found many ways to use them, with forestry and farming being just two examples. Recreation, conservation, water supply and **alternative energy** are some other examples of ways in which they can be used.

 THINGS TO DO AND THINK ABOUT

1. From the information on these pages, list five ways in which farmers can use upland glaciated areas.
2. Make a table to show the benefits and problems of forestry in upland glaciated areas. Try to think of five benefits and five problems.

 ONLINE TEST

Test yourself on farming and forestry online at www.brightredbooks.net/N5Geography

35

PHYSICAL ENVIRONMENTS
LAND USES: INDUSTRY

■ Working ● Inactive ▲ Restored

Diagram 1.47 Quarries in the Yorkshire Dales National Park.

VIDEO LINK

For more, watch the 'Quarrying in the Yorkshire Dales' clip at www.brightredbooks.net/N5Geography

ONLINE

Follow the 'Quarrying in Yorkshire' link to learn more at www.brightredbooks.net/N5Geography

DON'T FORGET

Quarrying is a suitable industry in the Yorkshire Dales because it provides precious employment in an area where it is scarce. Carboniferous limestone and gritstone are important rocks with many different uses.

DON'T FORGET

The worst effect of quarrying in protected landscape areas, such as the Yorkshire Dales, is the visual pollution or 'blot on the landscape'. There is little doubt that the presence of quarries has a negative economic impact on tourism in the surrounding areas (e.g. Horton-in-Ribblesdale).

QUARRYING IN THE YORKSHIRE DALES

The Yorkshire Dales **National Park** is famous for its carboniferous limestone scenery, both above and below ground. It has about 50% of all the limestone pavement found in the UK and is well known for its spectacular scenery of rolling hills separated by numerous dales. Around 9.5 million visitors arrive in the Yorkshire Dales National Park each year, spending over £400 million in the local economy. Industries, and in particular quarrying, also play a major part in the Dales economy, with at least nine working quarries situated within the National Park itself. Limestone is the main rock that is quarried but gritstone and shale are also extracted in some quarries.

Quarrying is an important industry, employing 7% of the workforce and adding some £6 million a year to the economy of the Dales. In a largely rural community of some 20 600 people, where the largest town is Settle with just over 2000 inhabitants, employment opportunities are scarce and jobs provided by quarrying are welcomed by most locals. There is a long history of quarrying and mining in the Dales and most of the existing quarries had been excavated before the area was designated as a National Park in 1954.

In Ribblesdale, north of Settle in the western Dales, there are four quarries at Horton, Arcow, Dry Rigg and Gigglewick, with a fifth near by at Ingleton. The employment opportunities provided by these quarries are vital in an area where farming is difficult and there are few other jobs outwith the tourist industry.

Horton Quarry: visual impact.

The carboniferous limestone has a variety of uses, including aggregate for the construction industry, where it is also used in high-grade concrete, flux for the steel industry, building stone, lime-based fertilisers for agricultural use, as a component in glass making and even in products such as toothpaste and cosmetics.

Millstone grit is used for surfacing roads, footpaths and airport runways.

QUARRYING AS A LAND-USE CONFLICT IN LIMESTONE AREAS

Quarrying is in direct conflict with the aim of the Yorkshire Dales National Park to conserve and enhance its natural beauty, wildlife and cultural heritage. Many visitors to the Yorkshire Dales simply cannot understand why such major industrial activity should be permitted within its boundaries. In other countries, such as Canada, industrial activity in a National Park would not be permitted. Horton Quarry in Ribblesdale is a significant excavation that is not easily missed by the many visitors to this area. In particular, over 100 000 walkers per year arrive in Ribblesdale to complete the Three Peaks Challenge, which involves walking between and climbing the mountains of Whernside, Pen-y Ghent and Ingleborough (the last being visible on the skyline in the photo). Yet it is recognised that the employment provided by quarrying is a valuable financial benefit to the local communities. There are several problems associated with quarrying in the Yorkshire Dales National Park:

- Visual **pollution** caused by quarries and also by the ugly industrial buildings used to process quarried rock.
- Noise pollution caused by blasting.
- Air pollution produced by dust following blasting.

contd

Physical Environments – Land uses: Industry

- Heavy goods vehicles (HGVs) cause traffic **congestion** on the narrow roads in the Yorkshire Dales: 75% of rock is transported by road.
- HGVs create additional noise and air pollution and may cause vibration damage to old buildings that are close to the roads.

Much has been done to alleviate these problems. The quarrying companies themselves have worked closely with local communities to identify ways in which the impact of the quarries can be reduced. Some of these measures are listed below:

- Quarry companies such as Lafarge Tarmac, Hansen Aggregates and the Pioneer Group hold regular consultative meetings with local communities.
- Quarry workings have been landscaped as far as possible, often by planting trees to screen them.
- Ugly quarry buildings have been removed from the skyline and, where possible, placed out of sight within the quarries themselves.
- Blasting is limited to times when there are fewer visitors, such as first thing in the morning and not on weekends and bank holidays.
- An increasing percentage of rock is transported by rail (e.g. from Swinden Quarry) to reduce the number of HGVs on nearby roads. At Swinden, this has resulted in over 150 fewer HGVs daily, or 42 000 fewer each year. Twenty-five per cent of all rock quarried in the Dales is now transported by rail.
- In Ribblesdale, Hanson Aggregates have agreed that HGVs from their three main quarries at Ingleton, Horton and Giggleswick will not pass through Settle, the main town in the area. Lorries are routed away from this community on to the main A65 route. This initiative has resulted in at least 2600 fewer lorries per year in Settle.
- It is common practice for HGVs to have tarpaulins covering their loads (this is called **sheeting**) and also for lorries to be washed down on leaving the quarry sites. Both of these measures help to reduce the amount of dust produced when transporting the rock.

A train loaded with rock leaving Swinden Quarry.

An HGV exits the wheel wash at Swinden Quarry.

Swinden Quarry: rock crushing buildings below quarry rim.

Solutions to the conflicts created by quarrying have met with a considerable degree of success. Local communities feel that they have a genuine say in developments at quarry sites and that their concerns are taken seriously. The development of rail facilities at Swinden Quarry has helped to significantly reduce the road congestion and pollution caused by HGVs. Tarmac were awarded a Biodiversity Benchmark Award at all three of their quarry sites (Swinden, Arcow and Threshfield) by the Yorkshire Wildlife Trust in recognition of their efforts to protect and promote wildlife habitats. Quarrying will always remain a contentious issue within our National Parks, but communities and quarry companies in the Yorkshire Dales continue to work together with the National Park Authority to minimise the environmental impact of this important industry.

 ONLINE TEST

Test yourself on industry online at www.brightredbooks.net/N5Geography

THINGS TO DO AND THINK ABOUT

1. Use Google Maps Street View to look at some of the quarries named on this page. The aerial shots will give you a good idea of the visual impact.
2. List the main conflicts caused by quarrying. Try to give examples of solutions for each conflict you mention.

PHYSICAL ENVIRONMENTS

LAND USES: WATER STORAGE AND SUPPLY, AND HYDRO-ELECTRIC POWER

WATER STORAGE AND SUPPLY

Northern and western parts of the UK have a water surplus. In other words, more rain falls than there is water required by the population. However, southern and eastern areas of the UK have a water deficit – there is insufficient rainfall to meet their water needs. Over many years, various upland areas of the UK, where there is high rainfall, have been developed to store and supply water. This can then be transferred to other areas where there is a need for water. With the construction of reservoirs in upland glaciated areas or river valleys the water is sometimes also used to generate hydro-electric power (**HEP**).

Examples of water transfer schemes in Britain are:

- Elan Valley (Wales) to Birmingham
- River Vyrnwy (Wales) to Liverpool
- Kielder Reservoir (Northumberland) to Newcastle, Sunderland, Middlesbrough and London
- Thirlmere and Haweswater (Lake District) to Manchester
- Loch Katrine to Glasgow
- Loch Lomond to Central Scotland and Edinburgh.

> **VIDEO LINK**
>
> Watch the 'Water supply infrastructure' clip for more information at www.brightredbooks.net/N5Geography

To develop water supply schemes large enough to supply whole cities requires a great deal of planning and engineering. Dams need to be constructed, pipelines put in place and if electricity generation is part of the scheme, then power stations must also be built. Inevitably this causes a great deal of environmental disruption and the development of water supply or hydro-electric schemes can be highly controversial. River valleys and glaciated U-shaped valleys are often suitable for water supply schemes as they are in, or flow from, areas of high rainfall; they have narrow valleys that can be easily dammed; and they often have a large area upstream that can be flooded for use as a reservoir. Areas with limited agricultural use and few other established land uses are particularly suitable for water supply and storage schemes.

Kielder Water dam.

Kielder Water

In 1975 work started on the Kielder Water Reservoir in Northumberland, which was designed to supply water to the conurbations of north-east England: Newcastle, Sunderland and Middlesbrough. By 1982 the reservoir was completed, creating an artificial lake some 9 kilometres long and holding over 200 billion litres of water. Kielder Water is an example of a river valley being used for a water supply and renewables scheme (HEP). The table below indicates some of the advantages and disadvantages of water supply schemes such as Kielder Water.

Kielder Water.

Advantages and disadvantages of hydro-electric power schemes

Advantages	Disadvantages
6 megawatts of electricity from hydro-electric power schemes. Multi-purpose development with tourists being encouraged to visit as well as water supply and electricity generation. Establishment of wildlife reserves and eight sites of special scientific interest (SSSIs). 200 billion litres of water stored to help with water supply in drier parts of England. The dam can help with flood prevention. Visiting tourists help to bring up to £6 million into the economy of surrounding settlements.	River flow downstream is altered, often with significantly reduced amounts of water. 58 families lost their homes as the reservoir filled up and flooded them and the entire village of Kielder was submerged. Other wildlife habitats were lost as the reservoir filled up. An area of outstanding natural beauty was lost as the reservoir filled up. Thousands of hectares of farmland were lost forever. Critics say that the Kielder Dam is visual pollution.

contd

Physical Environments – Land uses: Water storage and supply, and hydro-electric power

Coire Glas pumped-storage HEP scheme

Planning permission has been granted for the power company Scottish & Southern Energy (SSE) to construct a new pumped-storage HEP station at Coire Glas in the Laggan area, north of the Great Glen. This development will cost £800 million, including the construction of a dam that will create a new reservoir at Loch a Coire Glas, an underground power station and water tunnels with an outlet into Loch Lochy. A pumped-storage scheme allows water from the upper reservoir to be released to create power as it flows downhill at times of peak electricity demand. Electricity can then be transmitted to consumers via the National Grid. At times of low demand, power from the National Grid can be used to pump water back from the lower to upper reservoirs, in this case from Loch Lochy to Loch a Coire Glas. This ensures that there is plenty of water in the top reservoir to meet the next national peak in demand for electricity. The dam, at 650 metres long and 92 metres high, will be the largest in the UK and the electricity generated will be enough to power 1 million homes. The Coire Glas scheme is an example of **land-use conflict** in an area of upland glaciation.

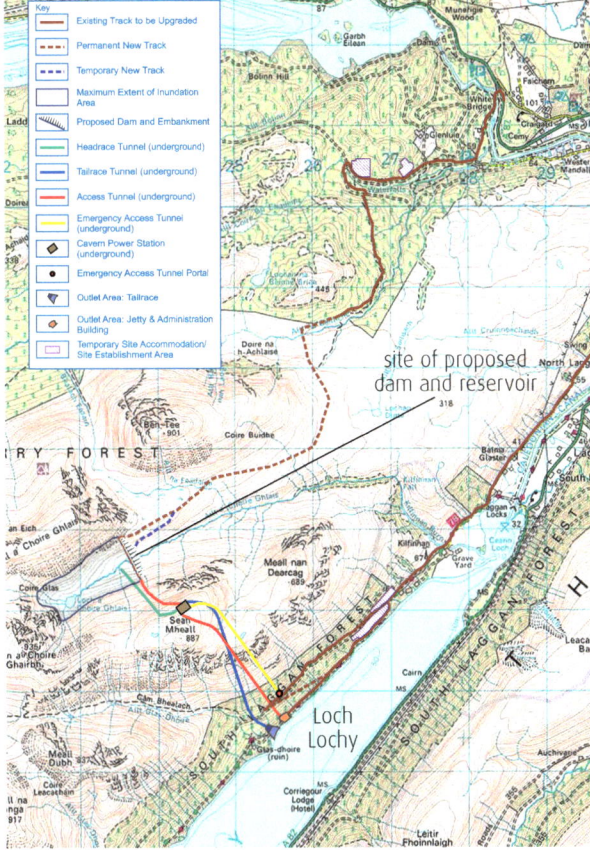
Diagram 1.48 Map of Coire Glas HEP scheme.

The advantages of the Coire Glas pumped-storage scheme are:

- the dam and reservoir will be concealed from view by the surrounding hills
- the power station and tunnels (for water) will be underground so visual disturbance will be minimal
- it is in an area of high rainfall
- the steep-sided glaciated valley is ideal for the construction of the dam
- its construction will help Scotland to meet its target of 100% of electricity coming from renewable sources by 2020
- it will provide 150 jobs during construction
- it provides a major environmentally **sustainable** energy source.
- a minimum of 12 permanent jobs will result from this development

The potential disadvantages of the Coire Glas pumped-storage scheme are:

- construction work could result in up to 25 lorries an hour on the main A82 road through the Great Glen in addition to the existing traffic levels
- there will be considerable environmental disruption and visual pollution during the construction phase
- this disruption could adversely impact local businesses who rely on tourism
- an area of wild and beautiful scenery will be altered forever
- wildlife habitats will be lost as the reservoir fills up and floods the valley.

In granting consent for the Coire Glas scheme in December 2013, the Scottish Government recognised that the disadvantages listed above are balanced by the benefits which the scheme will bring for UK energy security and in reducing carbon emissions.

Site of proposed dam and reservoir at Loch Coire Glas

DON'T FORGET

HEP is a type of renewable energy. It is sustainable and has zero carbon emissions once the infrastructure is built. It is ideal for a mountainous country with high rainfall. In Norway 98.5% of electricity is generated by HEP.

THINGS TO DO AND THINK ABOUT

Explain why the Coire Glas pumped-storage HEP scheme is suitable for an area of upland glaciation such as that of Laggan, north of Loch Lochy.

ONLINE TEST

Test yourself on water storage and supply online at www.brightredbooks.net/N5Geography

PHYSICAL ENVIRONMENTS

LAND USES: RENEWABLE ENERGY

The renewable energy industry in Scotland has been likened to the oil boom of the 1970s as developers and energy companies invest billions of pounds in sustainable energy schemes around the country. Hydro-electric power schemes (described on the previous page), wind farms, tidal power schemes and wave farms have all received huge investment as government grants encourage companies to develop these in order to reduce carbon emissions and to meet the government target of 100% of Scotland's electricity production being from renewable sources by 2020 in fact 97.4% was produced from renewable sources.

ONLINE
Check out the 'Scottish renewables' link for additional information at www.brightredbooks.net/N5Geography

DON'T FORGET
Wind power is especially suited to upland glaciated areas and coastal areas.

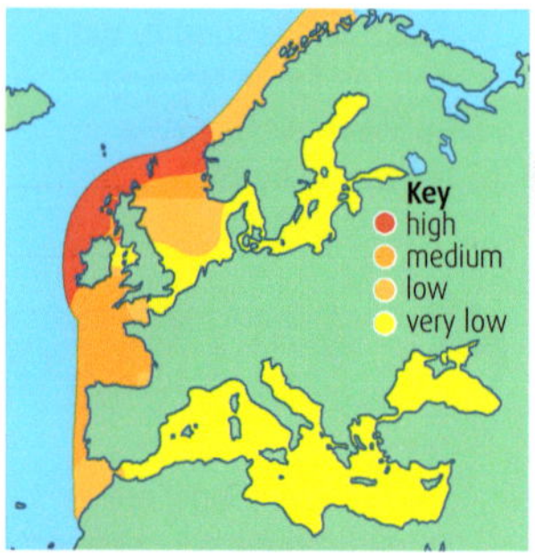

WIND FARMS

Over 25% of Europe's available wind energy crosses Scotland and its surrounding coastal areas, with 47% of the UK's wind power already being produced in Scotland. Wind farms are one of the 21st century's most controversial developments throughout the UK. Hundreds of wind farms have been constructed. The largest land-based wind farm in Europe is at Whitelees on Eaglesham Moor, 15 kilometres south of Glasgow. It has 215 wind turbines, with the capacity to provide electricity to power 300,000 homes. There are also 130 kilometres of tracks, built to construct and service the turbines, which visitors use for walking, mountain biking and riding.

However, the biggest potential for wind power is now seen to be in the development of offshore wind farms. A group of energy companies are constructing the world's largest offshore wind farm in the northern Moray Firth off the Caithness and Moray coastlines. Already partly operational, it is being developed by Moray Offshore Renewables and when complete, will consist of up to 339 turbines (depending on the height of the turbines) over 114 square miles of sea, within sight of coastal communities such as Wick, Helmsdale, Brora, Golspie and Dornoch. It is due to be completed by 2024, providing enough electricity to power up to over one million homes. Offshore wind farms are an example of conflict in the use of resources in coastal areas.

Developers claim that the construction of this offshore wind farm will create hundreds of jobs in the north-east of Scotland and that the £4.5 billion investment will boost the economy of this remote area. They also claim there will be minimal disruption to the sea bed as only the construction of foundations for the steel towers on the sea bed is required. The development will make use of Scotland's huge potential for wind power and help it to meet its requirement to cut carbon emissions.

However, there have been many objections to the proposal, including one from the Caithness Windfarm Information Forum. Objectors claim that the farm will be visible from a huge area of Caithness and Sutherland, and that this will have a detrimental effect on tourism, adversely affecting local businesses. The **RSPB** is concerned at the effect on sea birds in the area, while fishermen have expressed concern at the potential loss of fishing grounds. There are also fears that the wind farm will be a hazard to shipping, especially in stormy or misty conditions, as this is a busy area with ships travelling from the Atlantic through the Pentland Firth into the North Sea.

Diagram 1.49 Wave energy potential in Europe.

MARINE RENEWABLES: WAVE AND TIDAL POWER

Scotland's waters are estimated to have around a quarter of Europe's potential tidal energy resource and 10% of its potential wave resource. The **European Marine Energy Centre (EMEC)** in Orkney has a larger number of wave and tidal energy devices being developed and tested than anywhere else in the world. It is the world's first grid-connected, accredited facility for testing wave and tidal power machines. Recently, teams from Japan, South Korea, China, the USA and Canada have visited the site and

contd

Physical Environments – Land uses: Renewable energy

signed agreements to develop similar test facilities in their own countries with help and advice from EMEC. On land, next to the main wave test centre, is a small, landscaped, electricity sub-station where wave power is sent into the national electricity grid. The marine power companies also make use of purpose-built harbour facilities at Stromness, Kirkwall and Lyness on Hoy to berth and service their wave and tidal power machines. There are already hundreds of jobs in Orkney that depend on the marine renewables industry, with the prospect of many more as further developments take place.

Plans are in place to build enough wave and tidal farms in the waters around the coast of Orkney and the Pentland Firth by 2020 to generate 1.6 gigawatts of electricity – enough to power more than half a million homes all year round! One such project is the Meygen tidal energy farm planned for the waters of the Pentland Firth between Caithness and Orkney, which has one of the most powerful tidal streams in the world. Work on this site has already started and it is expected to be the world's first large-scale tidal energy farm with up to 269 turbines which will be constructed at and transported from the Nigg Energy Park on the Cromarty Firth.

There are concerns that wave and tidal farms could have a negative impact on bird and marine life. Surveys are currently being carried out to assess their impact on whales, dolphins and other marine species. Although wave and tidal power devices are offshore, there is still the need to connect them to the electricity grid and this requires the construction of onshore sub-stations and control rooms. To safeguard the machines, a marine exclusion zone is needed around them, adversely affecting fishermen and raising safety concerns for other boat users. Some disturbance to the sea bed is unavoidable in order to anchor the machines and also to connect sub-sea cables. There is also anxiety about the visual impact of wave farms close to the coast.

The marine renewables sector represents an important use of coastal areas. It is an industry still in its infancy, but Scotland has the unique potential to develop this resource. Despite financial and environmental concerns, there is little doubt that offshore wind, wave and tidal power will play a big part in generating energy for Scotland for many decades to come.

- Hundreds of jobs; prospect of many more
- Sea bed disturbance
- Construction of new harbour facilities
- Effect on birdlife and marine life
- Need for onshore sub-stations and control centres
- Orkney and Scotland known worldwide as leading centre of expertise on marine renewables
- Tidal energy is completely reliable
- Need for marine exclusion zones
- Effect on livelihood of fishermen
- Potential for huge amounts of clean energy
- Visual impact of wave forms
- Combats global warming

Diagram 1.52 Issues linked to marine renewables in Orkney.

 DON'T FORGET

Marine renewables, such as offshore wind, wave and tidal power, attract government and EU grants, as well as local investment from the companies involved, boosting the economies of coastal areas.

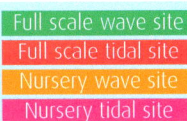

Diagram 1.51 Location of EMEC wave and tidal power test sites.

 DON'T FORGET

A recent study suggested that tidal energy turbines placed in the Pentland Firth between Orkney and the Scottish mainland could provide 43% of Scotland's electricity requirements.

THINGS TO DO AND THINK ABOUT

1. Construct a table to show the advantages and disadvantages of the Caithness offshore wind farm development.
2. Look at the diagram of issues linked to marine renewables. List the benefits and problems of this type of energy (there are six of each).
3. Find out more about wave and wind power via the EMEC link at www.brightredbooks.net/N5Geography

 ONLINE TEST

Test yourself on renewable energy online at www.brightredbooks.net/N5Geography

HUMAN ENVIRONMENTS

POPULATION: GLOBAL POPULATION DISTRIBUTION 1

AN OVERVIEW

At the end of 2021, the world's population was 7.9 billion and is forecast by the United Nations to reach 8 billion in 2023. However, although there is an ever increasing number of cities with over a million people, vast areas of our planet remain empty with few, if any, people. The distribution of people on our planet is measured by **population density**. This is calculated by dividing the population of a place or country by its total land area. For example Orkney has a total land area of 975 square kilometres and it had a population of 21 400 in the 2011 **census**, giving a population density of 21.9 persons per square kilometre.

Population density: $\frac{\text{total population}}{\text{total land area}}$

Population density of Orkney: $\frac{21,400}{975}$ = 21.9 people/sq.km.

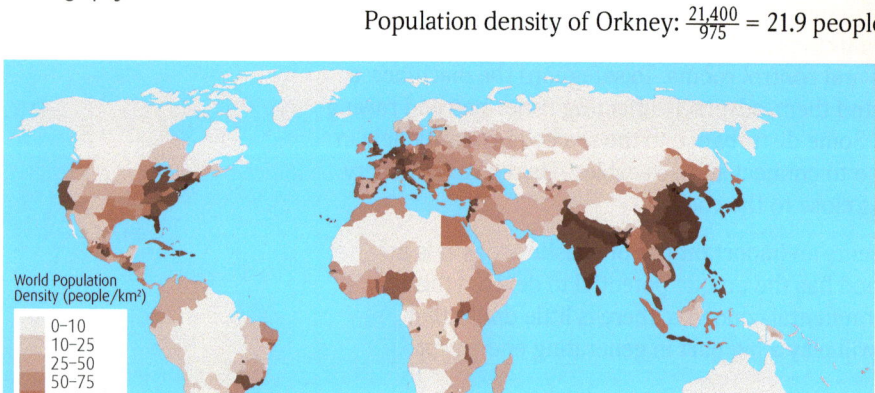

Diagram 2.1 World Population Density Map.

It is important to note that population density figures are averages and that within each area there may be large variations. In Orkney, for example, there are uninhabited islands where population density is 0, while in the main town, Kirkwall, there are areas with over 1000 people per square kilometre. Areas with few people are **sparsely** populated, while areas with many people are **densely** populated. This is shown on the world population density map.

ONLINE TEST

Take the test on population density at www.brightredbooks.net/N5Geography

ONLINE

Learn more about population density at www.brightredbooks.net/N5Geography

DON'T FORGET

Populations can be sparsely distributed or densely packed depending on a combination of both human and physical factors.

FACTORS AFFECTING POPULATION DISTRIBUTION

The world population density map shows that there are many heavily populated areas, the most densely packed country being Singapore with an average of 7 301 people per square kilometre (see table below). However, many cities are much more densely populated than this. Manila, capital city of the Philippines, has an average population density of 43 079 people per square kilometre and Paris has 20 741 people per square kilometre. Areas which are sparsely populated remain so because they have disadvantages which discourage people from settling there. Many of these are to do with physical factors such as the landscape or climate, but human factors such as **accessibility** or employment opportunities are also important (see diagram 2.2).

Most densely populated countries

Rank	Country	Population	Area (km²)	Population density (people/km²)
1	Singapore	5 183 700	710	7301
2	Hong Kong	7 061 200	1104	6396
3	Bahrain	1 234 596	750	1646
4	Bangladesh	152 518 015	147 570	1034
5	Taiwan	22 955 395	36 190	634
6	Mauritius	1 288 000	2040	631

contd

Human Environments – Population: Global population distribution 1

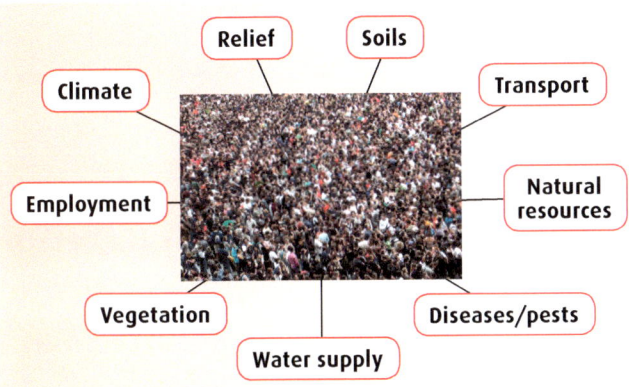

Diagram 2.2 Factors affecting population density.

Densely populated Hong Kong, China.

Climate

People like to live where the climate is not too extreme. Many of the most densely populated areas of the world have moderate climates: neither too hot and dry nor too cold and wet. Western Europe is an example of just such an area. The Netherlands, with its temperate climate, is the ninth most densely populated country in the world (404 people/km^2). Places with warm climates are pleasant to live in (such as coastal areas of the Mediterranean), but areas of tropical rainforest, for example, are too hot and humid, making it uncomfortable to live and work in. In some areas this has been overcome by introducing air conditioning. So Singapore, although it is only one degree north of the Equator, has become a very densely populated city nation (7 301/km^2). Other parts of the world, such as the islands of Svalbard in Norway, are quite the opposite. There are only three or four months when temperatures in Longbearbyen (78°N), the only large town in Svalbard, stay above freezing with average July temperatures of 4°C, while in mid-winter the average is only −17°C. Combined with the severe winter storms, this makes the climate here very harsh and inhospitable. It is therefore not surprising that Svalbard's population density is just 0.03 persons/km^2.

Relief

Flat land is easy to build on and so people favour flatter areas for settlements. The densely populated Netherlands exemplifies this too, where most of the country is flat and the highest point is only 323 metres above sea level. Mountainous areas are not only difficult to build on, they are usually colder, wetter and inaccessible too, so fewer people live there. Scotland's population distribution shows this well. Some 80% of the population lives in the Central Lowlands, while the rest of the country, especially the mountainous Highlands, is relatively sparsely populated (see map). There are some exceptions, however. In Kenya, the bulk of the population live in the Central Highlands where the altitude keeps the climate cooler and wetter and volcanic activity along the nearby African Rift Valley has resulted in particularly fertile soils for farming.

Soils

The availability of food supplies is a key consideration in the location of human population, and until comparatively recently this meant being close to a productive agricultural area. Flat or gently sloping land with good soil allowed the populations of nearby towns to maintain reliable supplies of food. Today, with rapid global transport systems, this is less of a consideration for many cities and developed world countries, but there is still a strong correlation between areas of high population density and good quality farmland. Mountainous areas are disadvantaged by a lack of good soil and available farmland.

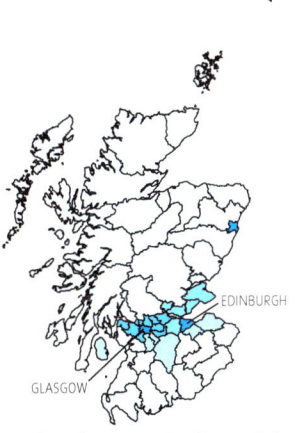

Most of Scotland's population is concentrated in the Central Lowlands.

 VIDEO LINK

Check out the 'Population density and distribution' clip at www.brightredbooks.net/N5Geography

 VIDEO LINK

Learn more about the population distribution of the UK by watching the clip at www.brightredbooks.net/N5Geography

 THINGS TO DO AND THINK ABOUT

Think about the area in which you live. What are the factors which have contributed to different population densities in the surrounding area?

HUMAN ENVIRONMENTS

POPULATION: GLOBAL POPULATION DISTRIBUTION 2

ONLINE TEST

Take the test on population density at www.brightredbooks.net/N5Geography

FACTORS AFFECTING POPULATION DISTRIBUTION (CONTD)

Transport

Accessibility is a key human factor affecting whether areas become densely or sparsely populated. Efficient transport systems allow goods to be traded easily and jobs to be created. Towns, which are route centres or ports, have often grown to become major cities because of their good transport links. A coastal location at the mouth of a river has been the reason for the rapid growth of many world cities. At these locations, goods can be traded overseas and shipments transferred to or from smaller vessels which can sail up river. Rotterdam at the mouth of the river Rhine/Maas, Liverpool (River Mersey), New York (Hudson River) and Shanghai (Yangtze River), are all examples of major world cities which have benefitted from this fortunate geographical advantage.

Liverpool, located at the mouth of the River Mersey

Natural Resources

Many settlements have developed solely to exploit natural resources which are close by. Historically, coalfields have had a major influence on the location of cities. Britain is a clear example of this. The most rapid urban development took place where industries were initially powered by coal mined nearby. Cities, such as Glasgow, Edinburgh, Newcastle, Manchester, Liverpool, Sheffield, Nottingham and Birmingham, grew as a result of rapid industrialisation made possible by easy access to nearby coal supplies. In Germany, the Ruhr area, with cities such as Dortmund, Duisburg and Essen, developed into a densely populated region for largely the same reason – proximity to supplies of coal and iron ore. However, not all settlements have been so successful. Some, such as those in the Canadian Rocky Mountains, were abandoned after the short-lived gold rush of the 19th century. There was no other reason to live there!

DON'T FORGET

The factors affecting population distribution in a particular place can change over time. For example the southern half of the island of Montserrat in the West Indies has become uninhabitable since 1997 due to dangerous volcanic activity, forcing the abandonment of its capital city, Plymouth. Climate change can also force populations to move away from land which has become unsuitable for farming.

Employment

In the twenty-first century it is possible to live in a remote location and work using a computer via the internet. However, for most people, being close to employment opportunities is still an important reason for relocating to major population centres. This is especially true in the developing world where **rural–urban migration** continues on a massive scale with improving employment still being one of the main reasons for it. **Urbanisation** in developed world countries, such as the UK, happened for the same reasons in the 19th and 20th centuries when people from the countryside moved to the cities to work in factories.

contd

Vegetation

Forested areas make settlement difficult. The tropical rainforests and the great coniferous forest areas of Scandinavia, northern Russia and Canada (**taiga**) are areas of sparse population. Dense forest also makes communications difficult, while climate in the rainforest is hot and humid and in the taiga the severe conditions in winter make these areas even less desirable to live in. Generally, many of the most densely populated areas in the world are areas of natural grasslands where soils are deep and fertile allowing productive farming to take place. Also, many densely populated areas have developed on areas which were once deciduous forest but have long ago been cleared to enable agriculture to take place and settlements to be built. Western Europe, including much of the UK, comes into this category.

Tropical Rainforest

Water Supply

The availability of water for domestic and agricultural use is a key factor influencing the distribution of world population. For this reason, the arid lands of deserts and semi-desert are sparsely populated.

Across North Africa, the vast Sahara is an area of very low population density. In the USA, massive engineering works have resulted in water transfer schemes which help keep desert cities alive. The Colorado River supplies several cities such as Los Angeles, San Diego, Las Vegas, Tucson and Phoenix which would otherwise not have been able to expand to their current size. Irrigated agriculture is also made possible by water taken from the Colorado River. Without these water supply schemes, these arid areas of southwest USA would be much emptier than they are today.

 ONLINE

Discover more about the factors affecting population density online at www.brightredbooks.net/N5Geography

Diseases/Pests

In some areas, water-borne diseases such as **malaria** have prevented areas from becoming more densely populated. Areas of marshland where insect pests, such as the mosquito, breed have often been avoided because of their association with high levels of disease.

Only by clearing and draining marshland can these areas become more suitable for permanent settlement.

Mosquito

 ## THINGS TO DO AND THINK ABOUT

Make a table with two columns. One for physical factors and one for human factors which affect population density. Study the factors affecting population distribution from the last four pages and place each into the correct column. Are there more physical or human factors?

 VIDEO LINK

Check out the 'Population video' at www.brightredbooks.net/N5Geography

HUMAN ENVIRONMENTS

POPULATION: SOCIO-ECONOMIC INDICATORS 1

AN OVERVIEW

A list showing the total populations of the world's five largest countries (and the UK) gives some basic information about them. However, in order to compare countries it might be more useful to have other information about their populations such as **population growth rate**, **life expectancy**, **birth rate** and **infant mortality**. This data allows more meaningful comparisons to be made about the **standard of living** within each country, taking into account, for example, the quality of health care as shown by life expectancy figures and levels of infant mortality. These figures, giving comparative information about countries' populations are called **social indicators**. This data is collected when governments carry out a **population census** – that is a statistical survey which gathers information about every individual in the country.

Table 1: Population information for the 5 largest countries and United Kingdom (data from CIA World Fact Book)

Country	Population	Population growth rate (% per annum)	Life expectancy (yrs)	Birth rate (per 1000 people)	Infant mortality (per 1000)
China	1 385 773 233	0.46	75.0	12.25	15.20
India	1 252 520 898	1.28	67.5	20.24	44.60
USA	320 114 481	0.90	78.6	13.66	5.90
Indonesia	249 940 303	0.99	71.9	17.38	26.06
Brazil	200 403 770	0.83	73.0	14.97	19.83
UK	63 395 084	0.55	80.3	12.26	4.50

Information about the wealth of a country, or its population, might include **gross domestic product (GDP)** per person, which shows the total value of the goods and services provided by a country divided by its population. Other useful comparative information about countries' economies might include data about value of imports and **exports** as well as the **percentage of the population working in agriculture**. These figures, giving information about a country's economy and therefore its wealth, are called **economic indicators**.

Table 2: Economic information for the 5 largest countries and United Kingdom (data from CIA World Fact Book)

Country	GDP	% workforce in agriculture	Exports ($ billion)	Imports ($ billion)
China	$9 100	34.8	2050	1817
India	$3 900	53	309	500
USA	$49 800	0.7	1612	2357
Indonesia	$5 000	38.9	189	179
Brazil	$12 000	15.7	242	239
UK	$36 700	1.4	481	646

When combinations of different social and economic information are viewed, the data is collectively known as **socio-economic indicators**. A socio-economic indicator is any type of data which gives an insight into either the population or economy of a country or both. Socio-economic indicators are often used to make comparisons between individual or groups of countries. As well as all of those listed above, individual socio-economic indicators might include, for example, the **number of patients per doctor**, **unemployment rate**, **average income** or the **percentage of the population with an internet connection**. When making comparisons between countries, the greater the number of socio-economic indicators which are considered, then the more accurate the conclusions are likely to be.

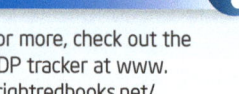

ONLINE

For more, check out the GDP tracker at www.brightredbooks.net/N5Geography

VIDEO LINK

Watch the explanation of GDP clip online at www.brightredbooks.net/N5Geography

contd

Human Environments – Population: Socio-economic indicators 1

A particularly useful socio-economic indicator is the **literacy rate**, which gives the percentage of the population that can read and write. Literacy rates give a clear indication about the effectiveness of a country's education system and also give an indication about a country's wealth, as the poorest countries cannot afford to run schools which are accessible to the whole population.

INTERPRETING SOCIO-ECONOMIC INDICATORS

Table 3 shows socio-economic indicators for selected developed and developing countries.

Table 3: Socio-economic indicators for selected countries (data from CIA World Fact Book)

Country	Population	Gdp	Life expectancy	Literacy rate	% Urban population	Access to clean drinking water
Bolivia	10 461 000	$5000	68	87%	67%	88%
Cambodia	15 206 000	$2400	63	78%	20%	64%
Germany	81 147 000	$39 100	80	99%	74%	100%
Japan	127 253 000	$36 200	84	99%	67%	100%
Netherlands	16 805 000	$42 300	81	99%	83%	100%
Sierra Leone	5 613 000	$1400	57	35%	38%	55%
Sweden	9 119 000	$41 700	81	99%	85%	100%
Uganda	34 758 000	$1400	54	69%	13%	72%
UK	63 395 000	$36 700	80	99%	80%	100%
Zimbabwe	13 183 000	$500	54	91%	38%	80%

By looking at the figures in Table 3, it is clear that the population figures do not help in determining a country's level of development. Uganda, for example, has a population of almost 35 million and a very low GDP, while Sweden has a much smaller population (9.1 million) and Japan a much larger one (127.3 million), yet their GDPs are much higher than Uganda's. However, there are much clearer correlations between some of the other figures, giving a much better impression of a country's level of development. The level of GDP in a country correlates directly with the life expectancy, the literacy rate and the percentage of the population with access to clean drinking water. This might be explained by the fact that wealthier countries can afford to invest in better education, good health services and improved water supplies for their people. Having access to clean water will also help to improve life expectancy as people will be affected by fewer infections and diseases from drinking contaminated water.

ONLINE TEST

Test yourself on socio-economic indicators online at www.brightredbooks.net/N5Geography

DON'T FORGET

Any measure of a country's society or population is a socio-economic indicator.

THINGS TO DO AND THINK ABOUT

1 Study the data in Table 3 and decide which countries are developed and which are developing. There are five in each category.

2 Compare the figures for life expectancy in Table 3. Which country has the highest figure and which two have the lowest? Now try to write down as many reasons as you can to explain why the figures are so different.

HUMAN ENVIRONMENTS

POPULATION: SOCIO-ECONOMIC INDICATORS 2

DISADVANTAGES OF SOCIO-ECONOMIC INDICATORS

As well as those mentioned on pages 46–47, Diagram 2.3 shows some further examples of socio-economic indicators.

average no. of years of schooling

patients per doctor

average no. of cars per family

% population with access to clean water

literacy rate

Gross Domestic Product (GDP)

infant mortality

average income

daily food intake

life expectancy

% population with internet access

population growth rate

Diagram 2.3 Selected examples of socio-economic indicators.

Blackburn vs St Albans

As with all national data, individual socio-economic indicators only give averages for the whole country and so it is important to be aware that these figures may not present a wholly accurate picture. There are several reasons for this.

Firstly, because each indicator is an average, it may disguise the fact that large variations may exist within a country. For example, literacy rates may be much higher in developing world cities where most children are able to go to school, than in **rural** areas where fewer schools exist and children may be required to work on their parents' farms.

Secondly, there may be big regional variations within individual countries. In the United Kingdom, for example, average income in Blackburn, Lancashire was just £15 700 in 2011 where as in St. Albans, Hertfordshire, it was virtually double at £30 446. Often levels of wealth might be higher close to the capital or other large cities, while in remote rural areas or depressed regional towns, poverty might be much more prevalent.

In developing world countries, there is often a hidden or **informal economy**, where people may barter or exchange goods and services without money ever changing hands. In this way, they may be able to get most of their requirements by growing their own food and exchanging it for other items they need. So, although they may appear to have a very low income, families are able to exist reasonably well without needing much money. For this reason, data about GDP or average income may be partially misleading.

Human Environments – Population: Socio-economic indicators 2

Also, a **development gap** exists between wealthy developed nations and poorer countries in the developing world. For example, a family in the United Kingdom may need well over £100 a week to be able to get their weekly shopping from the supermarket, whereas, in say, Tanzania, this amount of money might keep a family well fed for several months. As prices are much lower in developing world countries, an apparently small amount of money such as £5 might go a long way to feeding a family for a whole week, whereas in the UK, it would be completely inadequate. It is important to remember this when interpreting economic indicators such as GDP or average income.

COMPOSITE INDICATORS

Any one development indicator on its own does not give a clear picture of a country's level of development, so a number of composite development indicators have been introduced. The **Physical Quality of Life Index (PQLI)** combines **life expectancy**, **literacy** and **infant mortality rates**. This is calculated by giving each of the three indicators a score from 0 to 100, with 0 being the worst. The total for all three indicators is then averaged so countries have a PQLI index between 0 and 100.

A more commonly used composite indicator is the **Human Development Index (HDI)** which is calculated using **average income per person**, **average number of years at school** and **life expectancy**. The HDI was developed, and is still used by, The United Nations to gauge the levels of development in individual countries. Countries are given a score between 0 and 1, where a figure closest to 1 represents the most developed nation. In 2013 this was Norway with a figure of 0.955. One advantage of the HDI is that it combines social and economic indicators to give a more accurate picture of the standard of living within a country. It is still worth remembering that it is an average and wide variations may exist within a country.

Human Development Index 2014 (figures from United Nations Human Development Report 2015)

Rank	Country	Human Development Index score
1	Norway	0.944
2	Australia	0.935
3	Switzerland	0.930
4	Denmark	0.923
5	Netherlands	0.922
14	United Kingdom	0.907
184	Burundi	0.400
185	Chad	0.392
186	Eritrea	0.391
187	Central African Republic	0.350
188	Niger	0.348

VIDEO LINK
Watch the clip 'Human Development Index' at www.brightredbooks.net/N5Geography

ONLINE TEST
Test yourself on socio-economic indicators online at www.brightredbooks.net/N5Geography

DON'T FORGET
All socio-economic indicators including composite indicators are averages. Wide variations may exist between different regions or sectors of society within a country.

ONLINE
For more on economic and social indicators, check out the links at www.brightredbooks.net/N5Geography

ONLINE
Check out the BBC's analysis of their socio-economic survey 'The Great British Class Experiment' at www.brightredbooks.net/N5Geography

THINGS TO DO AND THINK ABOUT

1. What three measurements are used for the:
   (a) Physical Quality of Life Index?
   (b) Human Development Index?

2. Which measurement, the PQLI or the HDI, is most useful? Explain why.

3. The percentage of the workforce in agriculture is 1.7% in the UK and 55.6% in the Central African Republic. Explain what these figures show about the level of development in the Central African Republic compared with the UK.

HUMAN ENVIRONMENTS

POPULATION: BIRTH AND DEATH RATES 1

DON'T FORGET

A country's population growth rate also incorporates immigration and emigration. So the difference between birth rate and death rate in a country may not equate exactly to the population growth rate because of many people moving into or leaving a country. This is the case for Moldova where many people have emigrated.

ONLINE

Have a look at the 'Breathing Earth' link to see the world's population growth in real time at www.brightredbooks.net/N5Geography

VIDEO LINK

Check out the video '7 Billion, National Geographic Magazine' at www.brightredbooks.net/N5Geography for more on population growth.

AN OVERVIEW

The **birth rate** is the number of babies born per 1000 people in the population per year. The **death rate** is the number of people who die per 1000 per year. Both of these figures are **socio-economic indicators** and are useful when comparing population statistics and assessing a country's level of development. The average worldwide birth rate in 2015 was estimated at 18.7. This equates to 251 births per minute, or 4.2 births per second! The worldwide death rate in 2015 was estimated as being 7.9, which equates to 197 deaths per minute or 1.8 per second. The **natural population increase** can be calculated by subtracting the death rate from the birth rate. Often this is converted into a percentage figure, showing **population growth rate** per year.

Natural Population Increase = Birth rate – Death rate.

Natural Increase in World Population (2015): 18.7/1000 – 7.9/10000 = **10.8**/1000

Expressing this as a percentage gives the global population growth rate: **1.08%**

It is worth noting that, although this natural increase figure appears quite small, it represents in 1.08% annual increase in the world population of 7.3 billion. This is an extra 78.8 million people on the planet each year! There is considerable concern that this represents an unsustainable growth in the human population and that there will simply not be enough resources on the planet to go round if this rate of increase is not reduced. The United Nations and many national governments have therefore been introducing policies which will help to reduce birth rates in countries where these are high.

Country	Birth rate	Death rate	Population growth rate
Libya	18.74	3.56	4.85%
Zimbabwe	32.41	11.4	4.38%
Niger	46.84	13.07	3.32%
Uganda	44.5	11.26	3.32%
Burkina Faso	42.81	12.21	3.06%
Latvia	9.91	13.6	-0.61%
Ukraine	9.52	15.75	-0.63%
Estonia	10.38	13.65	-0.66%
Bulgaria	9.07	14.31	-0.81%
Moldova	12.38	12.61	-1.02%

Countries with high and low population growth rates
(Note that these natural increase figures also take account of migration and so may not represent the difference between birth and death rates.)

FACTORS AFFECTING BIRTH RATES

Birth rates may be high for many reasons. For example, cultural tradition and religion may encourage couples to have children. In some societies it is seen as a sign of virility for a man to father lots of children. In certain religions the use of contraceptives may be frowned upon. Poverty may encourage families to have lots of children so that they can be sent out to work and earn more money. Much work has been done by national governments and international organisations, such as the United Nations, to reduce the number of children being born. Some of these strategies are described below.

contd

Human Environments – Population: Birth and death rates 1

Family Planning Programmes

One of the most effective ways of reducing birth rates is by ensuring that access to contraceptives and family planning is as widely available as possible. In many developed world countries this is already the case but access to family planning services is limited for many people in certain developing world nations. When wider access is made available, people can then choose when and if they want to have children. Family planning services involve educating people about **contraception** as well as ensuring their easy availability. In Tanzania, where the average number of children per woman is 5.4, the government aims to increase the use of contraceptives and family planning advice from 27% of the population in 2010 to 60% of the population by 2015. Tanzania itself is a poor country and cannot afford to introduce such a programme on its own, but the United Nations Population Fund (UNPF) is contributing close to £1.5 million as aid to assist with this.

Better Health Care

Improving health care is a proven way of reducing birth rates. If people are healthier, they can work for longer and may perceive less need to have children for financial reasons. Reducing infant mortality in particular, will persuade parents that there is no longer any need to have extra children as a kind of insurance policy in case some of them die young. This does not have to involve the construction of new multi-million pound hospitals, but can be achieved by building extra clinics in less accessible rural areas, vaccination programmes and community education about reducing infection. This might be as straight forward as reinforcing the message about basic hygiene and hand washing.

ONLINE TEST
Take the birth and death rates test online at www.brightredbooks.net/N5Geography

DON'T FORGET
As well as census information from which birth and death rates can be calculated, UK law requires all births, marriages and deaths to be registered within 3 weeks. This keeps population information accurate and up to date.

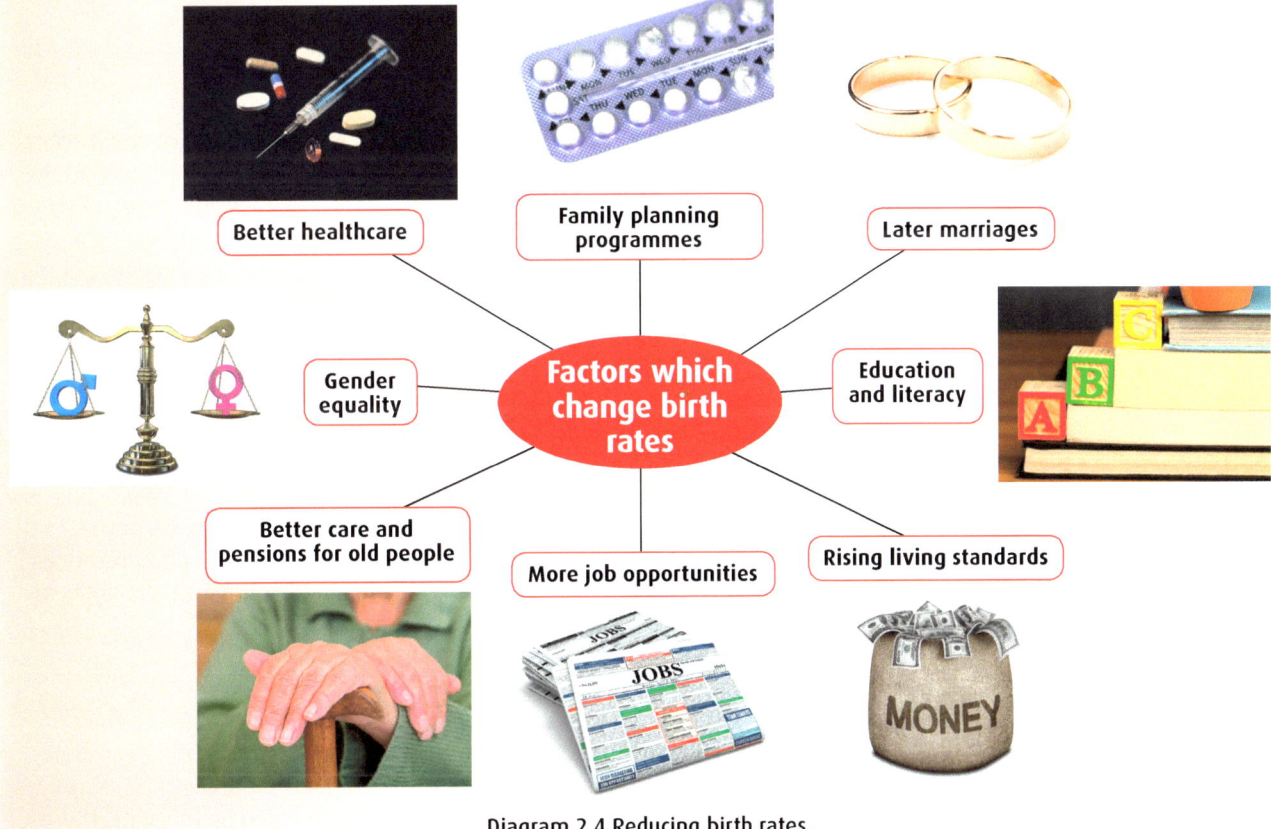

Diagram 2.4 Reducing birth rates.

THINGS TO DO AND THINK ABOUT

Calculate the natural population increase for:

(a) the United Kingdom (birth rate 12.26, death rate 9.33)

(b) the United States (birth rate 13.66, death rate 8.39).

HUMAN ENVIRONMENTS

POPULATION: BIRTH AND DEATH RATES 2

ONLINE

Read more about the rising average marriage age by clicking the link at www.brightredbooks.net/N5Geography

DON'T FORGET

"No country can really develop unless its citizens are educated." – Nelson Mandela

ONLINE

Find out more about the worldwide issue of equal access to education by clicking the 'UNICEF – Basic education and gender equality' link at www.brightredbooks.net/N5Geography

FACTORS AFFECTING BIRTH RATES (CONTINUED)

Later marriages

There has been a cultural shift in the UK in recent decades, with the average age at which people get married increasing steadily. In 1981 the average age at which women married was 23, while for men it was 25. Now, the average for women is 30 and for men it is 32.

This is a trend which has been repeated in many developed world countries. One of the consequences of this is that, on average, couples then have fewer children. Few countries have deliberately raised the legal age of marriage to reduce births, but in China a minimum age of 24 was introduced for marriage, although this has now been reduced to 22 for men and 20 for women.

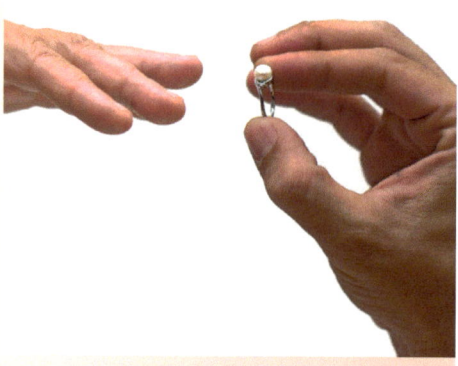

The average age for men and women in the UK to marry has increaed by seven years since 1981.

Education and Literacy

Ensuring that girls and boys, even in the remotest areas of a country, have equal access to education helps to increase literacy and also the aspirations of young people. If they receive a good education and are able to apply for well paid jobs, a whole range of new opportunities becomes available, whereas previously marriage and child bearing at a young age might have been the only option for girls. If good jobs are accessible, people may choose to concentrate on their careers and have fewer children later on in life.

Gender Equality

There are many cultural barriers to achieving true equality. Ultimately, if girls have the same opportunities as boys, they may have the option of being able to pursue careers instead of becoming mothers at a young age. Along with access to good education and family planning services, girls are also more likely to be able to make informed choices about their lives. Working against this are cultural influences in some parts of the developing world where girls may marry at the age of 12 (under certain circumstances in Tanzania) and where girls may be denied access to education altogether (in areas of Afghanistan under Taliban rule).

Better Care and Pensions for Old People

In many societies, older people are traditionally looked after and cared for by their children. Having several children might increase the chances that one of them will be able to care for their parents in old age. By ensuring that pensions are sufficient for elderly people to live independently and that good care is provided by the state, there may be less need to have extra children for this purpose.

Rising Living Standards

Where standards of living have improved, people receive better wages and may perceive less need to have children who can be sent out into the community as wage earners. There is also a link between more affluent societies and lower birth rates which has been attributed to the desire to save for consumer goods such as cars or for foreign holidays. In order to do this, couples may make a conscious choice to have fewer children, as raising a child can be expensive.

Human Environments – Population: Birth and death rates 2

CHINA'S ONE-CHILD POLICY

The best known example of a country which introduced policies aimed at reducing birth rates is China. In 1979 the Chinese government decreed that couples may only have one child. The reason for such a drastic policy was a realisation that if the Chinese population continued to increase at such a rapid rate, food supplies would eventually start to run out and living standards would plummet. The one-child policy was rigorously enforced and women who became pregnant for a second time faced unrelenting pressure from the authorities to have an abortion. Couples who went ahead and had a second child had to pay back all the benefits that the first child had received and also faced the prospect of losing their jobs. Today, the Chinese government says that its policy has prevented 400 million births and that China is therefore a wealthier nation. There were some exceptions to the one-child policy. Couples in rural areas whose first child was a girl were allowed to try for a second child, while if a couple were both single children themselves, they were allowed a second child. At the start of 2016, the Chinese government introduced a two-child policy for the whole country. In 2021 the limit was increased to three children per family due to fears about China's very low birth rate.

VIDEO LINK

If you'd like to learn more about China's One-Child Policy, watch the documentary at www.brightredbooks.net/N5Geography

Chinese poster advocating its one-child policy.

DEATH RATES

Many of the factors described in this section as affecting birth rates have a similar impact on death rates. Improving living standards, better health care, improved care and pensions for the elderly are all examples of policies which have helped to bring down death rates. In a society with higher living standards, life is less physically demanding for many than it once might have been. This, together with improved medical care, helps to increase life expectancies and so lower death rates.

ONLINE TEST

Take the birth and death rates test online at www.brightredbooks.net/N5Geography

ONLINE

Follow the 'Factors affecting population growth' link at www.brightredbooks.net/N5Geography to explore this topic further.

THINGS TO DO AND THINK ABOUT

Japan's birth rate is 8.23 and its death rate is 9.27. There is no significant immigration or emigration. What are the likely consequences for Japan of its birth and death rates?

HUMAN ENVIRONMENTS
POPULATION PYRAMIDS

A **population pyramid** shows the **population structure** of a country. The population structure of a country is the relative proportions of males and females in each age group. Age groups are usually 0 to 4, 5 to 9, 10 to 14 years and so on. Population pyramids give a very good visual representation of a country's population structure.

DEPENDENCY RATIO

The combined proportions of a country's population that are under 15 and 65 or over are said to be the dependent population – that is children and retired people. The age groups in between, that is from 15 to 64, are the economically active population. This is the section of the population that is mostly working and whose taxes help to fund services for the dependent population such as schools, clinics, pensions and old people's homes. For a country's economy to be healthy, it is important that the economically active sector of the population is a reasonably large proportion of the total and, conversely, that the dependent sector is reasonably small. The dependency ratio is a useful socio-economic indicator and is usually expressed as a percentage:

$$\text{Dependency Ratio} = \frac{\text{no. of people aged 0 to 14} + \text{no. of people aged 65 or over}}{\text{no. of people aged 15 to 64}} \times 100$$

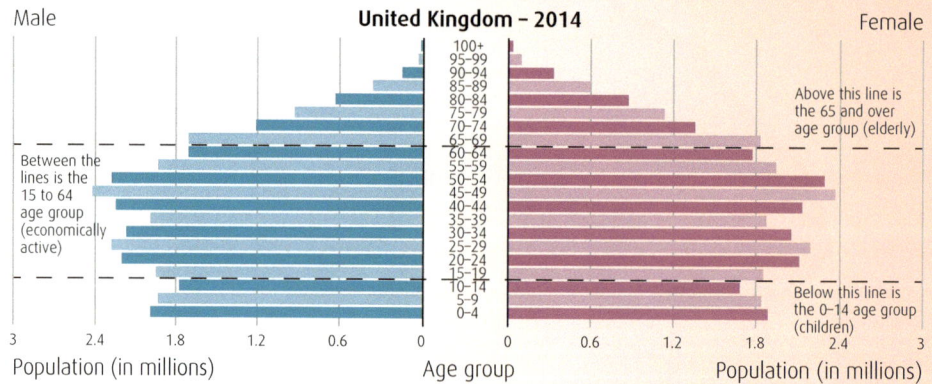

Diagram 2.5 Population pyramid for the United Kingdom (2014).

COMPARING DEVELOPING AND DEVELOPED COUNTRIES

The **population pyramid** for the United Kingdom (above) shows the **population structure** based on the latest population census data. The UK is a **developed country** and so there are a lot of middle aged and elderly people because life expectancy is quite high.

The population pyramid for Tanzania (Diagram 2.6) is typical of a **developing country** with a high birth rate (37 per 1000) and relatively low life expectancy (61 years). There is a very wide base and just under half (45%) of the population are under 15. At the top end of the graph, only 3% of the population is over 65.

The population pyramid for Germany is quite typical of a developed world country with a low birth rate (only 8 per 1000) and high life expectancy (80 years). There is a narrowing base indicating that the birth rate has fallen in recent years and there is a much smaller proportion (13%) of the population under 15 than in Tanzania. In complete contrast, at the top end of the graph, there is a much bigger proportion (21%) aged 65 and over, reflecting the long life expectancy and high living standards. Of concern to the German government is the large number of people who will reach retirement within the next 20 years and the falling number of births. Combined, these two factors could result in a very high dependency ratio, as there will be fewer workers and a very high proportion of older people.

contd

ONLINE

Try this out for yourself by following the 'Global Population' link at www.brightredbooks.net/N5Geography

DON'T FORGET

A population census is a survey of a country's population characteristics carried out by its government. This is where the data for population pyramids comes from.

Human Environments – Population pyramids

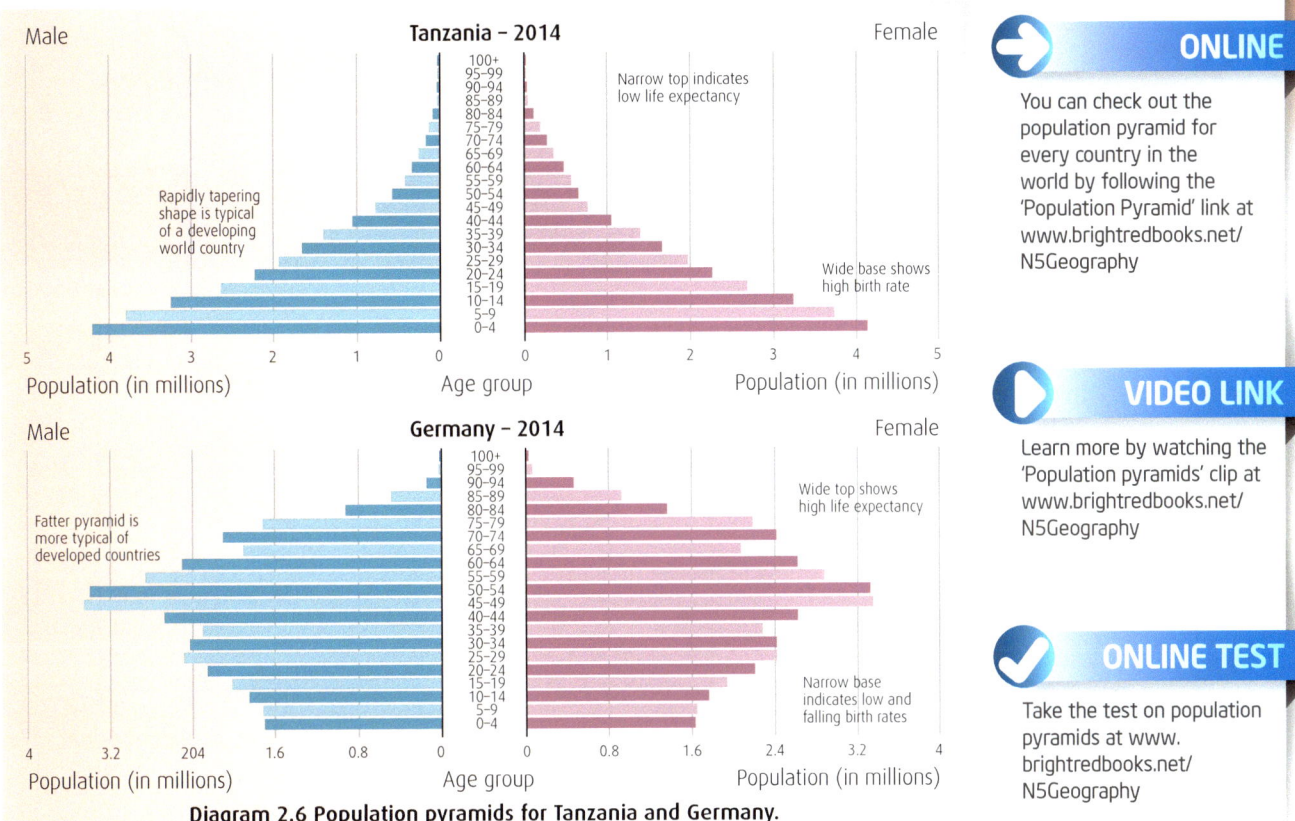

Diagram 2.6 Population pyramids for Tanzania and Germany.

AGEING POPULATIONS

In Japan (Diagram 2.7) they are already dealing with the scenario of an ageing population. This is the consequence of widespread reduction in birth rates and rising living standards. Due to a combination of low birth rates, good health care and high living standards, almost 25% of Japan's population are now aged 65 or over. This proportion is predicted to rise to 38% by 2055. One effect of an ageing population is that the total population is likely to fall, as fewer children are born. The under-15 age group in Japan accounts for only 13% of the total. The proportion of the working age population is also likely to shrink, resulting in a higher **dependency ratio**, where there are fewer working people to pay taxes which can go towards the needs of the retired people and children. This may result in a need to increase taxes, and raise the age of retirement. This is already happening in the UK where the retirement age is soon to be 67, and eventually 69, for both men and women. Another consequence of ageing populations is the need to spend significantly greater sums on health care, as ill health is more widespread among older people. Already in Japan there has been a surge in the number of businesses targeting the "grey" market by providing services such as residential homes and specially selected holidays.

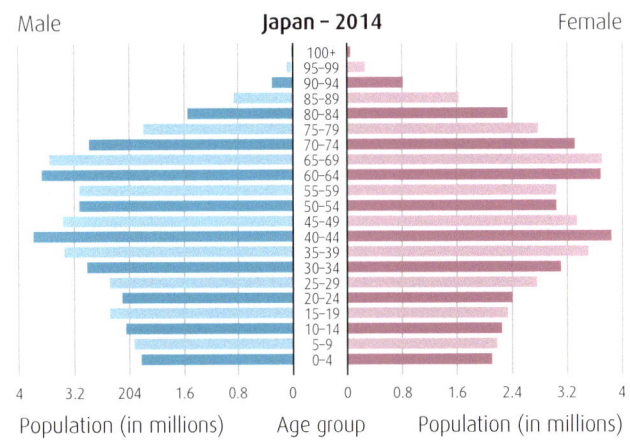

Diagram 2.7 Population pyramid for Japan.

> **ONLINE**
> You can check out the population pyramid for every country in the world by following the 'Population Pyramid' link at www.brightredbooks.net/N5Geography

> **VIDEO LINK**
> Learn more by watching the 'Population pyramids' clip at www.brightredbooks.net/N5Geography

> **ONLINE TEST**
> Take the test on population pyramids at www.brightredbooks.net/N5Geography

THINGS TO DO AND THINK ABOUT

1. What are the main differences between the population structures of developing and developed world countries? Try to give reasons for these differences.
2. Explain what difficulties a government in a country with an ageing population might face.

HUMAN ENVIRONMENTS
DEMOGRAPHIC TRANSITION MODEL

STAGE ONE

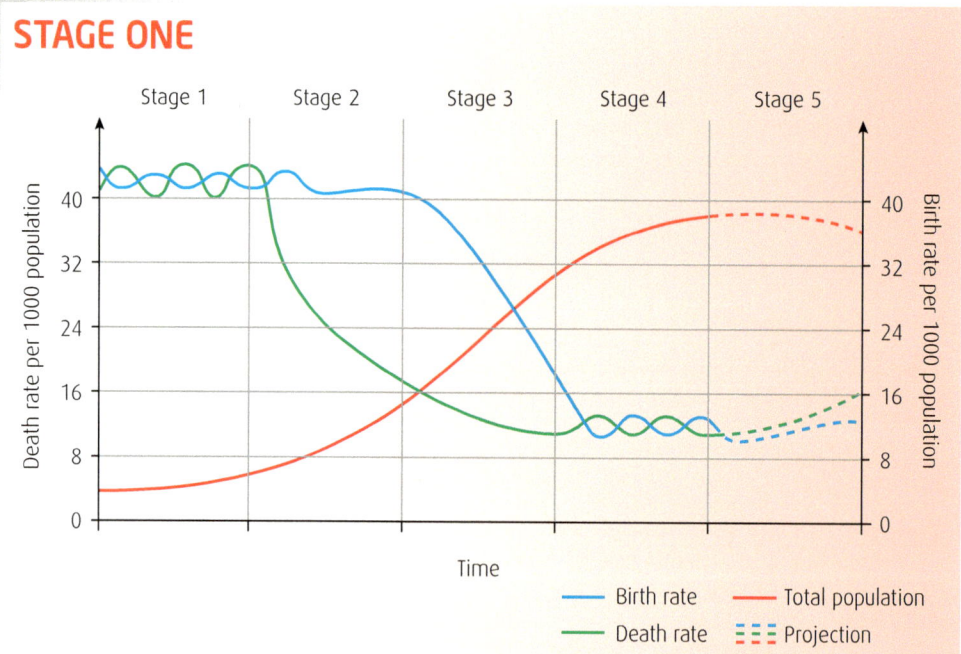

Diagram 2.8 Demographic transition model.

The **demographic transition model** shows how population alters over time as a result of changes to birth and death rates. In stage one, both birth and death rates are high, although the death rate fluctuates more than the birth rate, due to epidemics and famine for example. However, the total population does not change much as there is little natural population increase. Family sizes are large but infant mortality rates are high and life expectancy is low due to disease, famine and very challenging living conditions. This may have been the situation in the UK before the industrial revolution started in the mid eighteenth century.

STAGE TWO

In stage two, there is a dramatic drop in the death rate. This could be due to the introduction of clean water supplies, proper sewage systems and improved medical care for example. Birth rates stay high though because it takes time for people to realise that traditionally large family sizes are no longer necessary because of reduced infant mortality. The overall effect of this change is to cause a rapid increase in the total population, due to the large difference between the birth and death rates.

STAGE THREE

In stage three, the birth rate begins to drop, perhaps due to improved access to family planning and contraception, or a realisation among couples that there is less need to have so many children because of improvements to health care and living standards. However, it remains substantially higher than the death rate and so the overall effect is still a rapidly increasing population although the rate of increase is beginning to slow. The death rate continues to fall due to improvements in health care and living standards. Today, the population characteristics of many developing world countries fit this stage of the model. This gives cause for optimism about future global population growth which might rise more slowly and eventually stabilise.

Human Environments – Demographic transition model

STAGE FOUR

By stage four the fall in the birth rate has caught up with the falling death rate and so overall population increase slows or stops. Many countries in the developed world, including several in western Europe such as France and Spain, have population characteristics similar to this stage of the model.

STAGE FIVE

There is also a fifth stage which projects what may happen in the future as birth rates continue to fall because couples are having fewer children later in life. Also, death rates may rise slightly due to the dramatically increasing elderly population. This would give an overall reduction in population. In Europe, the population characteristics of Sweden and Italy fit this stage, while Japan's population profile also has many similarities to this stage of the model. China's one-child policy was introduced with the specific aim of not only stopping the country's rapid population increase but also significantly reducing the overall population. If the policy is maintained, China's population will fall as it enters stage five of demographic transition.

SUMMARY

The demographic transition model is useful in trying to understand the changing characteristics of global population. As society evolves and birth and death rates change, the model allows us to explain the changes and make predictions about future population size. The model may be less useful for predicting and understanding the demographics of some countries however, because it doesn't take account of international migration. The policy of allowing free movement of labour within EU countries, for example, has resulted in large population gains due to migration in countries such as Britain and Germany, with substantial drops in population in some eastern European countries such as Lithuania. This migration distorts the population patterns which otherwise might closely follow those of the demographic transition model.

ONLINE TEST

Take the test on the demographic transition model at www.brightredbooks.net/N5Geography

DON'T FORGET

The demographic transition model simulates what happens to a country's population over time. However, it is only an estimate – every country is different and its pattern of population growth may vary from that shown in the model.

ONLINE

Look at the CIA World Fact Book for further up-to-date information about population at www.brightredbooks.net/N5Geography

THINGS TO DO AND THINK ABOUT

Look at the data for each country in this table of birth and death rates. Copy it and complete it by estimating which stage of the demographic transition model the data fits best. The first one has been done for you. They all fit stages 3, 4 or 5.

Country	Birth rate	Death rate	Stage of demographic transition model
Finland	10	10	4
Serbia	9	13	
Tanzania	37	8	
Mali	46	14	
Japan	8	9	
Spain	10	9	
Jamaica	18	7	

HUMAN ENVIRONMENTS

URBAN GEOGRAPHY: URBAN ZONES

VIDEO LINK

Learn more about urban land use models by watching the video at www.brightredbooks.net/N5Geography

DON'T FORGET

Edinburgh and Leith were originally separate towns but as Edinburgh expanded outwards it merged with the port of Leith which now forms part of Edinburgh's inner city.

AN OVERVIEW

An urban land use model simulates the way in which different types of land use cluster together in cities. So for example, major services, such as theatres, council offices and department stores, are usually found in the most accessible part of a city (the centre) where they will be available to the highest number of people. The most common land use model has concentric land use rings radiating outwards from the centre. As the city grows outwards, many of the newest developments take place on the edge of the city where there is space and where land is cheaper. The cost of land in different urban zones is an important factor determining the land use in that zone.

Diagram 2.9 Model of urban land use zones.

CENTRAL BUSINESS DISTRICT (CBD)

Features of a city centre: Edinburgh.

The **central business district (CBD)** is the busiest part of a city. It is usually the place where the city first developed so historic buildings, such as Edinburgh Castle, are often found here. As this is the most accessible point of the city, there is high demand from businesses and services which want to locate here. This increases the cost of land and rents, so it is expensive to buy or rent property here. A typical CBD will have bus and train stations, cinemas, theatres and conference halls, shopping centres and department stores, restaurants, bars and clubs, as well as many public and private offices. Due to the high price of land, buildings often have a lot of floors to fit in more space and so skyscrapers are common. Shopping is one of the main purposes of a CBD as it is the most accessible point for customers.

INNER CITY

Inner city and suburban housing compared.

Inner city Dundee.

The **inner city** is a zone containing many older buildings which developed at a time when the city was growing rapidly. Housing was built quickly and cheaply to accommodate the many workers who moved into the city to work in factories. Much of the inner city developed before the advent of mass transport systems, so factories and housing were constructed close together. In Scotland tenements grew around the factories or shipyards, while in England terraced housing was more common. Often streets were built in a grid pattern with rows of parallel streets intersected by others running at right angles to them. This street pattern allowed the maximum number of buildings to be constructed in each area and is known as a grid-iron or **rectilinear** street pattern. The inner city today is characterised by tightly packed housing with a high population density. Land prices are still comparatively high, so there is little open space. There are few gardens or driveways and cars are parked on the streets. Inner cities have many old churches, small businesses and factories, pubs, small shops and supermarkets as well as schools to serve the many families who are resident here.

Human Environments – Urban geography: Urban zones

THE SUBURBS

Usually the **suburbs** are the largest urban zone, containing the bulk of a city's population but with a lower population density than in the inner city. Towards the edge of the suburbs, housing is newer, reflecting a city's outward growth over time. Suburban land prices are lower than those in the CBD or inner city as the suburbs are further away from the centre and so less accessible. This means that individual houses can be larger (but not necessarily cheaper) than in the inner city and most have gardens and garages so that cars are not parked on the streets. In more modern suburban areas, the road layout was deliberately curved with many cul-de-sacs or "dead ends". This street pattern is said to be **curvilinear** and prevents traffic from travelling along residential streets, making them both safer and quieter. In the 21st century, the suburbs are characterised by modern housing with double-glazed windows, central heating, gardens and garages. There is much more open space, often with tree lined streets, school playing fields, parks and golf courses, resulting in a more pleasant, quieter and less polluted **environment**.

Suburban housing: Dunfermline.

THE RURAL-URBAN FRINGE

At the edge of the city is the **rural–urban fringe** where the city meets the countryside. Often this is not a clear dividing line but a mixed area of new urban developments and farmland. Around many large British cities there is a **green belt** – an area with strict planning regulations designed to stop outward expansion, or **urban sprawl**, and to protect the countryside. However, in recent decades these planning restrictions have been relaxed and many areas of green belt land on the rural–urban fringe have seen urban development taking place. Sometimes this might be housing estates but there are many other land uses too. Land in the rural–urban fringe is desirable because:

- land is cheaper
- there is plenty of space
- it is undeveloped and so easier to build on than sites which have been used before
- there is a cleaner environment
- often there is good access, particularly if there is a city bypass or outer ring road
- modern transport networks mean people can commute easily.

Apart from housing, other developments might include out-of-town shopping centres, industrial estates and business parks. Leisure developments such as sports centres, golf courses, country parks and garden centres are located here too. Beyond the rural–urban fringe is open countryside but there may be villages close by which have grown rapidly and become urbanised because of their proximity to a city. Old houses have been modernised and new housing has increased the size of the village as **commuters** flock to buy houses here, far enough away from the city to escape some of the disadvantages, such as pollution, but still close enough to work in the city and benefit from its many services. Most cities are surrounded by several commuter villages or settlements.

 ONLINE

To learn more about suburbs, read up by following the links at www.brightredbooks.net/N5Geography

 ONLINE TEST

Take the test on urban zones at www.brightredbooks.net/N5Geography

The rural–urban fringe: Edinburgh.

 ## THINGS TO DO AND THINK ABOUT

1. The urban transect (Diagram 2.10) shows land use zones from the centre to the edge of the city. Copy the transect and use the information on these pages to complete it by adding labels for the inner city, suburbs and rural–urban fringe.

2. Look at the pictures of the inner city and suburban houses. Write down a list of differences and try to give reasons for each one.

3. Study the locations of Scotland's main cities (e.g. on Google Maps). Make sure you can locate them accurately on a blank map of Scotland. You should know the exact locations of Glasgow, Edinburgh, Dundee, Perth, Aberdeen and Inverness.

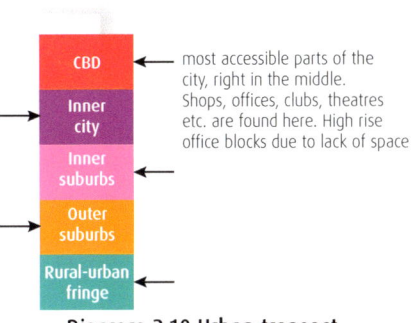

Diagram 2.10 Urban transect.

HUMAN ENVIRONMENTS

URBAN ZONES ON O.S. MAPS

Each urban land use zone has its own characteristics which are clearly recognisable on Ordnance Survey (O.S.) 1:25,000 and 1:50,000 maps. It is important to be able to identify the main road types correctly as they are a key feature of each land use zone. Diagram 2.11 shows some of the more common transport symbols used on 1:50,000 maps.

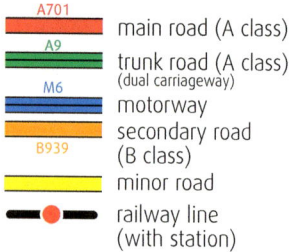

Diagram 2.11 Key transport symbols on 1:50,000 O.S. maps.

ONLINE

Check out the OS Maps key link at www.brightredbooks.net/N5Geography

DON'T FORGET

Only roads shown in blue are motorways! If you are describing roads on an O.S. map, be accurate or you will lose marks.

CENTRAL BUSINESS DISTRICT

As city centres developed before cars existed, their road pattern is often unplanned, containing many tightly packed small streets. All CBDs can be recognised by the large number of main raods and often railway lines which converge on them. This is because CBDs are route centres and the most accessible point of the city. Other key evidence which helps to identify a particular area of a map as the CBD includes the presence of many transport terminals such as bus and railway stations. Symbols for public services, such as museums, art galleries, tourist information centres and town halls, represent evidence of the CBD too. Often there will be a cathedral and/or a high number of churches which are clearly visible on the map.

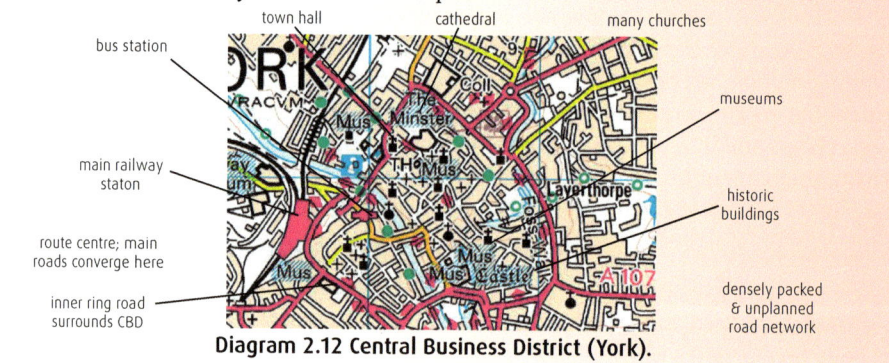

Diagram 2.12 Central Business District (York).

INNER CITY

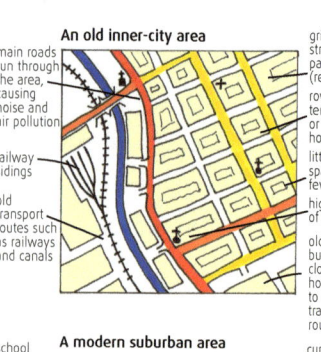

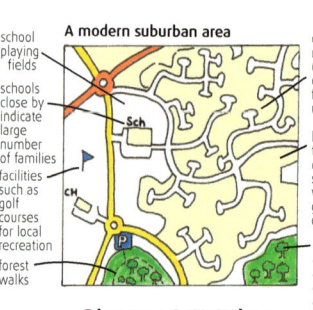

Diagram 2.13 Urban road patterns.

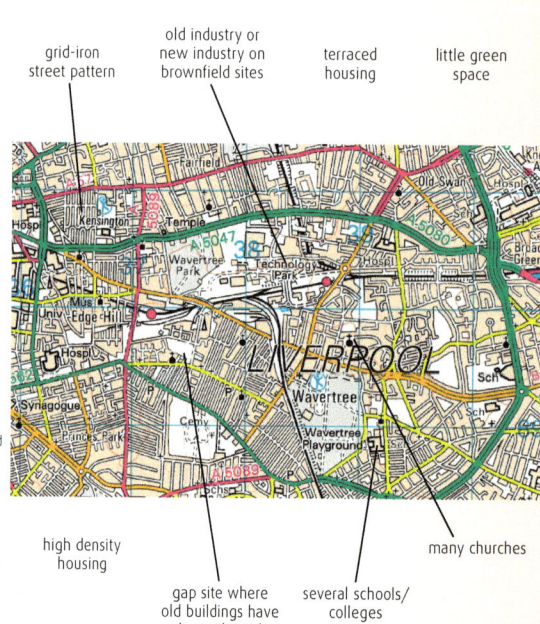

Diagram 2.14 Inner city Liverpool.

The inner city is essentially an old area and there is map evidence which will confirm this. Built in order to cram the maximum amount of housing into the smallest amount of space, the grid iron road pattern is easy to identify. Housing will be terraced, or tenement blocks in Scotland. There is little open space and there is often industry found among the residential areas close to old transport routes such as railway lines or canals. Industrial buildings stand out as large and often irregular in shape, sometimes with the symbol "works", possibly abbreviated to "wks." next to them. This is a busy part of the city and there may be many A and B class roads leading towards the CBD.

Human Environments – Urban zones on O.S. maps

THE SUBURBS

Towards the edge of the city in the more modern suburban areas, curvilinear street patterns are in evidence. These cul-de-sacs and crescents indicate the presence of modern detached or semi-detached housing complete with gardens and driveways. There is much more open space than in other parts of the city as shown by the presence of parks, sports centres, school playing fields, golf courses and even patches of woodland. There are fewer main roads amongst the housing areas than in the inner city, as they are designed to be safer, quieter and less polluted.

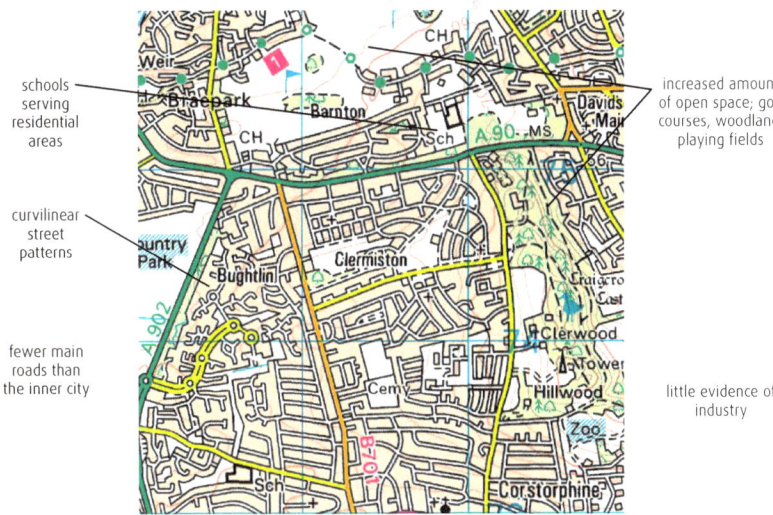

Diagram 2.15 Modern suburbs (Edinburgh).

THE RURAL-URBAN FRINGE

At the edge of the city there is a mix of countryside and modern urban developments. These can be new housing, trading or business parks, out-of-town shopping centres, leisure facilities such as golf driving ranges and sports grounds or country parks, created for city dwellers to be able to enjoy the freedom of open spaces and the countryside. There will be working farms interspersed among the new development and often evidence of newly-built urban bypasses or ring roads. There may be newly-built **park and ride** facilities to encourage motorists to leave their cars at the edge of the city and use public transport. Just beyond the city edge there may be commuter towns and villages which have expanded rapidly as the city has grown. Although new housing areas will have curvilinear road patterns, the rural–urban fringe appears much less regimented than either the inner city or the suburbs. This is because it is an area that is still rapidly evolving.

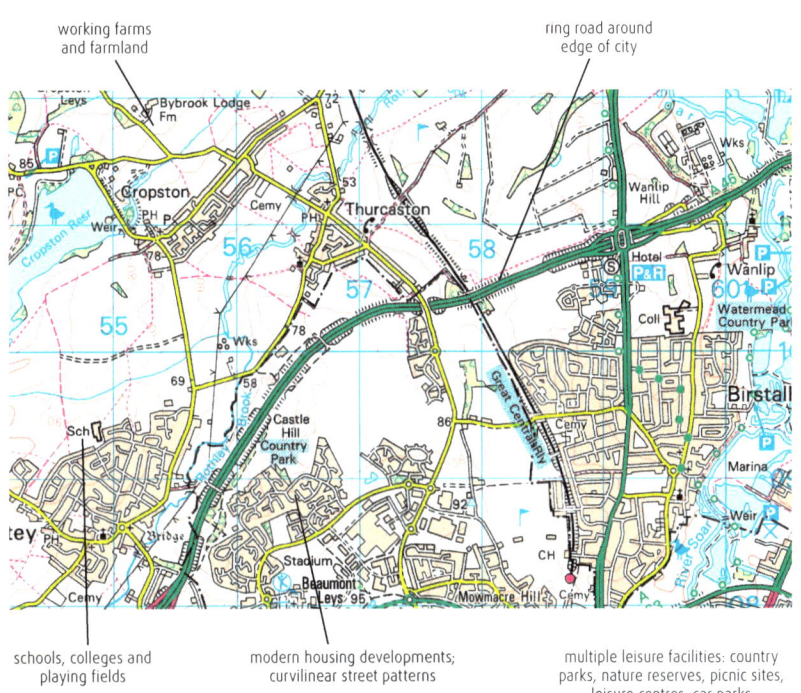

Diagram 2.16 The rural–urban fringe (Leicester).

THINGS TO DO AND THINK ABOUT

1. Make a list of the main identifying features when looking at each of the urban zones on Ordnance Survey maps.
2. Use Streetview on Google Maps to explore different types of urban area. Look at the types of building and compare the environments for the amount of open space, traffic, parked cars etc.
3. Learn to correctly identify the different road types. You should be able to name each different colour of road.

ONLINE TEST

Take the 'Urban zones on OS maps' test at www.brightredbooks.net/N5Geography

VIDEO LINK

Find out more about Ordnance Survey by watching the clip at www.brightredbooks.net/N5Geography

61

HUMAN ENVIRONMENTS

URBAN GEOGRAPHY: CHANGES IN THE CENTRAL BUSINESS DISTRICT

DYNAMIC CBD

> **DON'T FORGET**
>
> You should already know the main land uses in a typical CBD: offices, shops, transport hubs, entertainment venues, restaurants, fast food outlets.

The central business district is the most important part of a city. People travel to the CBD for work, shopping and leisure activities. It is often also the oldest part of town, and buildings or whole areas sometimes need to be renewed to bring them up to date and to meet the needs of the modern city. Also, as CBDs face increasing competition from out-of-town shopping centres and business parks, city councils want to modernise the CBD to keep it attractive for businesses, shoppers and visitors alike. City centres are **dynamic** – they don't stay the same for long as they are always changing.

CHANGE IN THE CBD

Some of the most common changes found in the CBD are:

- *Pedestrianisation* – shopping streets are made traffic free to make them safer and more pleasant for customers.
- *Covered shopping centres* – air conditioned, clean, weather proof with easy access for families and disabled people. CBDs have to compete with out of town retail parks.
- *Public transport* – is improved to give people an alternative to their cars (e.g. trams in Edinburgh).
- *High rise office blocks* – making the most of land which is expensive due to shortage of space for development.
- *Multi-storey car parks* – to get parked vehicles off the streets.

> **DON'T FORGET**
>
> Edinburgh's CBD is unusual in that there are many people who live there. Many old commercial buildings have been refurbished and converted to flats and new car-free apartments have been built (e.g. at Tollcross).

EDINBURGH CITY CENTRE: A CASE STUDY

Edinburgh is a good example of a developed world city. As the capital city of Scotland, home to 495 000 people, the second largest financial centre in the UK and home of the world's largest international arts festival, Edinburgh's CBD is vitally important in countless ways. Like most CBDs, the centre of Edinburgh has experienced massive change and continues to do so.

Historic skyline above Waverley Station in Edinburgh's CBD.

In 1995, the centre of Edinburgh was declared a United Nations World Heritage Site in recognition of the unique character of its buildings such as Edinburgh Castle, Holyrood Palace and the Georgian New Town. This has been a major reason why there has been little development of skyscrapers in Edinburgh – to protect the unique views of Edinburgh's historic buildings. However, there has still been lots of change in its CBD.

Transport

Edinburgh has worked hard to find ways of reducing the number of cars in the city centre. In 1992 a city bypass was completed so that vehicles no longer needed to drive through the CBD if they didn't need to be there. This helped to reduce vehicle numbers. Multi-storey car parks have been constructed, such as at Greenside, so that fewer cars need to park on the streets, thereby reducing congestion. Most important of all has been Edinburgh's policies to encourage people to travel to the CBD by using public transport. Greenways have been marked on most main roads into the centre. Greenways are bus priority lanes (painted green) which only buses, taxis and bicycles can use. A new bus station

Edinburgh tram.

contd

Human Environments – Urban geography: Changes in the central business district

was completed in 2003 with a new upmarket shopping mall (Multrees Walk) constructed above it. New stations have been built on the edge of the city, such as South Gyle and Edinburgh Park, giving people new ways to commute to and from the city centre. Meanwhile, cars have been banned from Princes Steet which is open only to trams, buses and taxis. All of these developments have been aimed at reducing the number of people who bring their vehicles into the city, so that congestion and air and noise pollution are reduced.

Edinburgh Quay: A new development on the edge of Edinburgh's CBD.

Old and New Together

Although there are no skyscrapers, there have been many other modern developments in the centre of Edinburgh. Land is **expensive** in any CBD, especially in a major city such as Edinburgh, so developers will make the most of any land they can get hold of. An old railway yard on the western edge of the CBD which was used as a temporary car park has been developed into a new financial district for the city along with a new 5-star hotel. On the south eastern edge of the CBD, land which was once used by the brewing industry has been transformed by two modern buildings – the Dynamic Earth Exhibition and the iconic building housing the Scottish Parliament. These modern buildings stand side by side with some of the oldest buildings in the city – the tenements at the foot of the High Street and Holyrood Palace. Preserving historic buildings and protecting the view of them has presented city planners with real difficulties. The Princes Mall underground shopping centre just north of Waverley Station was built below street level so that the unique view of Edinburgh's historic skyline, including the Castle, was protected. More than most cities, Edinburgh has imposed strong planning constraints on new city centre developments because of the importance of tourism to the city economy – Edinburgh Castle, for example, is the most visited tourist attraction in Scotland. Ensuring that the city's historic views are not diminished is good for business!

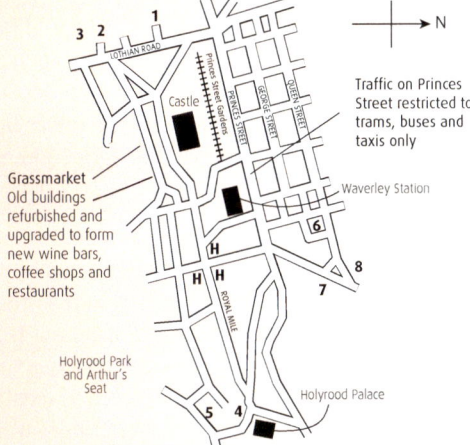

H Old buildings converted to new hotels (e.g. former Scotsman newspaper office)
1 Edinburgh International Financial Centre (former railway yard)
2 Edinburgh International Conference Centre
3 Edinburgh Quay leisure and office development
4 Site of Scottish Parliament (old brewery)
5 New Scotsman newspaper offices & Dynamic Earth exhibition
6 New shopping centre, new department store, new bus station
7 Omni Centre: new leisure complex & Greenside multi-storey car park
8 Redevelopment of the St. James Shopping Centre to be completed by 2021

Diagram 2.17 Sketch map showing selected changes in Edinburgh's CBD.

Developments, such as the new financial district at Clydesdale Plaza, the Dynamic Earth and Scottish Parliament buildings at Holyrood, the Omni centre and multi-storey car park at Greenside, and the new Edinburgh Quay leisure complex at the terminus of the Union Canal, are examples of how a CBD not only changes but grows outwards, absorbing land that at one time was part of the inner city. So it is possible for the location of the CBD to gradually change and grow with time.

ONLINE
Check out the website about Edinburgh's business sectors and explore the key areas of Edinburgh's CBD at www.brightredbooks.net/N5Geography

DON'T FORGET
Edinburgh's CBD & inner city continues to change with developments such as the opening in 2021 of the St. James Quarter shopping centre and the extension of the tram network to Leith and Newhaven (to be completed by 2023).

DON'T FORGET
Tourism in Edinburgh is hugely important to the city's economy, so all new developments are either directly for the benefit of tourists, such as new hotels, or are designed to blend in with Edinburgh's existing architecture and the historic skyline which tourists expect to see.

ONLINE TEST
Check out how much you know about changes to CBDs online at www.brightredbooks.net/N5Geography

THINGS TO DO AND THINK ABOUT

1 Study the information on these two pages, then close the book and try to write down six changes which have happened in Edinburgh's CBD.

2 Use the Ordnance Survey's Getamap or Google Street View websites to study the CBD of Edinburgh and find some of the sites described above.

HUMAN ENVIRONMENTS

URBAN GEOGRAPHY: CHANGES IN THE INNER CITY

Inner city problems: derelict properties.

Inner city problems: waste ground in Belfast.

Glasgow's high rise flats.

End of an era: demolition of tower blocks.

SLUMS AND SKYSCRAPERS

A number of problems are often associated with the inner city zone. Housing is old, densely packed and often of a poor standard. There are derelict or run-down properties, both residential and industrial, as well as patches of waste ground, creating an unappealing visual landscape. Inner city areas have been associated with high levels of unemployment, crime and even social unrest, as seen in the riots in some British cities in the summer of 2011. Traffic volumes, air and noise pollution can be problematic as vehicles travel to and from the nearby CBD.

To tackle the problem of poor housing conditions in the 1960s and 1970s, **urban redevelopment** schemes were implemented. High rise tower blocks were constructed to replace inadequate terraced or **tenement housing**, which often lacked basic amenities such as inside toilets or bathrooms. At first, the high rise developments appeared to be a good solution to some of the housing problems of the inner city. Areas of slum housing were cleared and replaced with tower blocks which were able to absorb much of the high density population of these areas. They were cheap and could be built relatively quickly. Each flat had modern amenities such as bathrooms, hot running water and central heating. More open space could be provided around the base of the tower blocks, opening up previously enclosed areas. In Glasgow, some 240 tower blocks were constructed.

Within just a few years however, it became evident that tower blocks were not the solution many people had hoped for. Badly constructed flats and poor ventilation resulted in problems with condensation. Elderly people often felt trapped in their flats by a lack of neighbourliness, contrasting sharply with the high levels of community spirit which had existed in the areas from which they had been moved. There were high levels of crimes such as muggings and vandalism, in the passageways and entrance halls. Lifts often broke down. There was a lack of places for children to play safely. In short, for many people the high flats were a disaster.

In the twenty first century, increasing numbers of tower blocks in the inner city have been demolished to be replaced by housing that is seen as more socially and visually acceptable. Low rise buildings, often with their own garden space, or tenement style buildings with secure entrances and space for residents' cars have been constructed in many inner city areas. There has been more emphasis on providing services, such as shops, leisure facilities and health care, within the inner city to promote a greater sense of community spirit. The provision of job opportunities is a key part of regenerating inner city areas and so there has been an emphasis on the creation of small trading estates or business developments.

contd

Human Environments – Urban geography: Changes in the inner city

THE REDEVELOPMENT OF LEITH

Redevelopment of the inner city often now involves modernising old buildings, including sometimes changing their use altogether. **Brownfield sites** are redeveloped and brought back to life. In Leith, old abandoned whisky warehouses at Commercial Quay have been brought back into use as modern flats with cafés and restaurants at ground level. Existing tenement buildings have been gutted, modernised and made fit for the twenty-first century. Where it has been necessary to demolish older buildings, they are often replaced with modern-style tenements designed to blend in with existing ones. Abandoned industrial land in Leith has been brought back into use as housing, shopping or leisure developments. The Ocean Terminal Shopping Centre is a major retail and leisure development in Leith designed to bring people back into the inner city. It contains over 70 shops including a major department store, coffee bars, restaurants, health and fitness suites, a cinema and over 1600 free car parking spaces. It is well connected by frequent public bus services and is home to the Royal Yacht Britannia where over 250 000 visitors a year pay to see round the Queen's former ship.

For many people, the inner city is an attractive and convenient place to stay. It is close to the workplaces, shopping, leisure and entertainment facilities of the CBD. Rent and property prices are affordable and modernised and updated housing developments provide people with the standard of accommodation they expect in twenty first century Britain. Since the 1980s, Leith has enjoyed something of a revival with the redevelopment of old industrial sites as small commercial premises or as affordable housing. The waterfront in Leith has attracted several upmarket restaurants and the area has become popular again, with boats moored on The Water of Leith in use as night clubs and the once polluted banks of the river, cleaned up and redeveloped as a public walkway.

In 1994 the new Scottish Office headquarters was built on the waterfront, bringing hundreds of new jobs to the area.

In the future, Forth Ports Authority plans to infill part of the Western Harbour area in Leith Docks and provide further new housing developments. By 2020 it is planned that the whole harbour area will provide up to 17 000 new homes.

The regeneration of Leith as an inner city area has brought about a transformation in its fortunes. Once seen as an area of sub-standard housing, derelict buildings and high unemployment, the image of this part of Edinburgh's inner city is now much more positive. To achieve this has required massive investment in new housing together with a deliberate policy of bringing this part of the city back to life by providing shops and services and, above all, employment opportunities. Leith, as with other inner city areas, will continue to evolve and adapt to social and economic change.

New developments in the inner city: Ocean Terminal.

Whisky warehouse converted to flats: Commercial Quay, Leith.

New tenement flats designed to blend in with old tenements.

VIDEO LINK
Watch the clip 'Local government: town planning and regeneration' for more on the demolition of Glasgow's towerblocks at www.brightredbooks.net/N5Geography

ONLINE
Learn more about the redevelopment of Leith online by following the link at www.brightredbooks.net/N5Geography

DON'T FORGET
Most construction in the inner city takes place on **brownfield sites**: land that has been built on before.

DON'T FORGET
Urban redevelopment in the inner city involves not only upgrading the buildings, but also providing adequate local services and employment opportunities for the population.

VIDEO LINK
For more, watch the STV news clip 'Urban regeneration project for Leith to be unveiled' at www.brightredbooks.net/N5Geography

THINGS TO DO AND THINK ABOUT

1. Make a list of the problems which are typical of inner city areas.
2. How has urban redevelopment in inner city areas changed since the construction of tower blocks?
3. Use Google Maps Street View to explore Leith, or any other inner city area, to see the new developments. Start by typing in Commercial Quay, Leith.

ONLINE TEST
Check out how much you know about changes in the inner city online at www.brightredbooks.net/N5Geography

HUMAN ENVIRONMENTS

URBAN GEOGRAPHY: CHANGES ON THE RURAL-URBAN (RURBAN) FRINGE

THE RURBAN FRINGE OF EDINBURGH

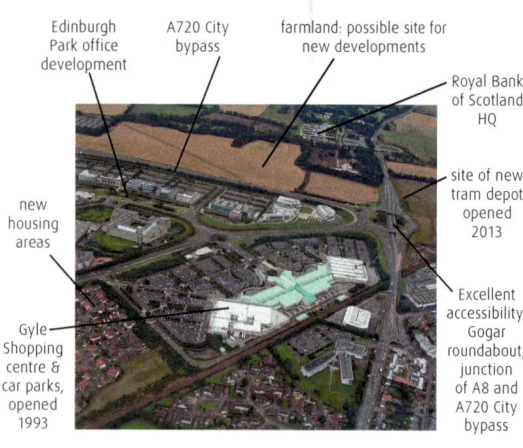

Edinburgh's rurban fringe: Gyle Centre.

Edinburgh is Scotland's capital city; it is the second largest city in Scotland (with 495 000 inhabitants) and home of the Scottish Parliament. With the exception of London, it is the biggest financial centre in Britain, has the two most visited tourist attractions in Scotland (Edinburgh Castle and The National Galleries of Scotland), each with over 1 million visitors per year and is home to the world renowned International Festival in the summer. Edinburgh is surrounded by many other settlements such as Dunfermline, Musselburgh, Dalkeith, Haddington, Tranent, Broxburn, Livingston, Newtongrange, Gorebridge and Penicuik. The combined total population of Edinburgh together with the surrounding towns is approximately 850 000. It is hardly surprising, therefore, that the city's growth is exerting considerable pressure on the surrounding greenbelt land and that developments on the rural–urban fringe continue to happen at a spectacular rate.

EDINBURGH'S WESTERN EDGE

New development on the rurban fringe: Gogar Tram Depot.

In 1992 the A720 Edinburgh City Bypass was completed around the southern edge of the city, leading from the A8 in the west to the A1 and Musselburgh in the east. Controversial at the time for its construction on green belt land around Edinburgh, this ring road now carries thousands of vehicles per hour and has been an important development in helping to reduce the amount of traffic in Edinburgh's CBD. This is one of the most common uses of land on the **rurban fringe** of major cities.

Since its construction, many new developments on the edge of the city have been constructed on **green field sites** in the gap between the edge of the city and the A720. This process is sometimes called **infilling** – where rural land between a city's edge and a new ring road is developed, expanding the city right up to the new road.

Following the construction of the City Bypass, the first major development in this part of Edinburgh's rurban fringe was the Gyle Shopping Centre, which opened in 1993. It employs some 2000 staff in over 70 different retail premises including a department store, major supermarket, coffee shops and restaurants. At the same time, new housing estates were completed nearby, as well as a new station on the Edinburgh to Fife railway line, allowing people to commute without using a car.

Close by, lies Scotland's largest business park – Edinburgh Park, where the first office was opened in 1995. Over 7000 people are employed here in a wide variety of different businesses. In 2003 a further new station was opened to serve the business park on the main Edinburgh to Glasgow railway line. Both the Gyle Centre and Edinburgh Park are also connected to Edinburgh Airport and the city centre by the new tram system.

Just beyond the City Bypass, lies the global headquarters of the Royal Bank of Scotland at Gogarburn. Officially opened in 2005, this woodland campus office development was designed to offer a unique working environment and facilities for its 3000 strong labour force. Contained within the building are shops, conference facilities, a health and fitness suite complete with swimming pool and medical centre.

VIDEO LINK

Check out the 'Commonwealth Games' clip at www.brightredbooks.net/N5Geography for a look at the development of this part of Glasgow's rurban fringe.

DON'T FORGET

Developments on the rural-urban fringe take place mostly on **green field sites**, that is land that has not been built on before. Usually, this is easier to develop than a **brown field** site.

contd

Human Environments – Urban geography: Changes on the rural–urban (rurban) fringe

Other developments in this rapidly changing part of Edinburgh's rural–urban fringe are the new tram depot at the Gogar roundabout and the Hermiston Gait Retail Park, just to the south of Edinburgh Park. Beyond the RBS village is Edinburgh Airport, a rapidly growing development on Edinburgh's western edge, employing over 2500 people. Next to the airport is the Ingliston **park and ride** facility where people can park their cars free and travel by bus into the city centre, helping to reduce traffic. There are now six park and ride facilities on the edge of the city with a further one at Ferrytoll just north of the Forth Road Bridge.

Apart from the obvious advantages of space to build and cheaper land, the developments in this area have several major transport advantages making it highly **accessible** for both workers and customers. These are:

- next to Gogar roundabout junction of A8 and A720 City Bypass
- next to M8 motorway linking Edinburgh to Glasgow
- close to the M9 and A90 roads leading to Stirling, central scotland, Fife and the north
- nearby Edinburgh International Airport provides important business links
- new stations at South Gyle and Edinburgh Park provide the option of rail travel
- Edinburgh's new tram system connects this area to the airport as well as the CBD.

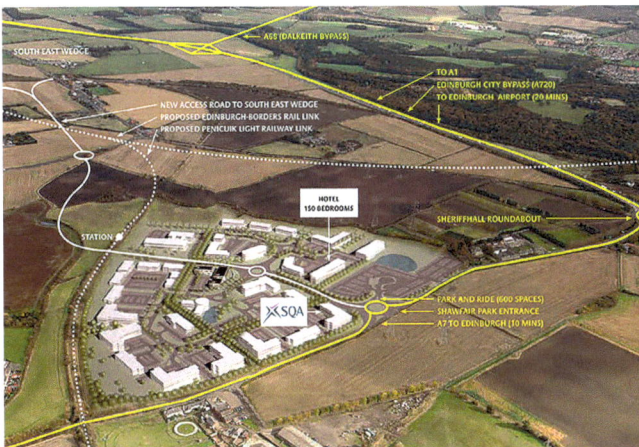

Shawfair Business Park and south-east wedge development.

Other Developments on Edinburgh's Rurban fringe

Some areas on the rural–urban fringe have limited potential for expansion. Immediately to the south of the city lies the Pentland Hills Regional Park, established in 1986, an important area for conservation, walking and outdoor sports, including the recently redeveloped snow sports centre and dry ski slopes at Hillend. Here, the City Bypass runs close to existing housing on the inside and is separated from the Pentland Hills by a number of golf courses, such as Torphin Hill, Swanston and Lothianburn on its outer edge. Altogether there are at least 10 golf courses on the western, southern and eastern sides of Edinburgh's rurban fringe. These, together with the Pentland Hills Regional Park, create a green barrier preventing further urban development towards the south.

There are other out-of-town retail parks on the southern edge of the City Bypass at Straiton Park where major retailers, such as Ikea, Costco and DIY superstores, act as magnets for shoppers, and also at Fort Kinnaird inside the bypass on the eastern edge of the city. Bilston Glen **industrial estate**, immediately south of Straiton, was built on the site of a former colliery.

Accessibility is a key factor for new developments in the rurban fringe.

A more recent development to the south-east of the city is the Shawfair Business Park and Sheriffhall **park and ride**. This is the first stage of a much larger proposed development which will see the creation of a new town with some 4500 homes built on, or close to, land formerly used by the Monktonhall Colliery. Included in this development will be a new station on the Borders railway line, which opened in September 2015. All of these developments represent further infilling between the edge of Edinburgh and the City Bypass.

 ONLINE TEST

Check out how much you know about changes on the rural–urban fringe online at www.brightredbooks.net/N5Geography

 THINGS TO DO AND THINK ABOUT

1. Write down at least six advantages for developers of land on the western edge of Edinburgh's rural–urban fringe.

2. Copy the table and match the developments on Edinburgh's rurban fringe with the correct place name from the list below. The first has been done for you.

   Hermiston Gait; Gogar; Straiton Park;
   Fort Kinnaird; Gyle Centre; Shawfair;
   Pentland Hills; Sheriffhall;
   Edinburgh Park; Gogarburn; Ingliston.

Rurban fringe development	Place name
1. Airport park & ride	Ingliston
2. Shopping centre (opened 1993)	
3. RBS headquarters	
4. Tram depot	
5. Regional Park	
three other shopping centre	
6.	
7.	
8.	
9. Scotland's largest business park	
10. Business park south-east of city	
11. Park and ride scheme next to this business park	

HUMAN ENVIRONMENTS

URBAN GEOGRAPHY: CITIES IN THE DEVELOPING WORLD

VIDEO LINK

Check out the 'Learn about contour lines' clip online at www.brightredbooks.net/N5Geography

SHANTY TOWNS

Brazil, with a population of 213 million (2021), is the sixth most populous country in the world. It is also the fifth largest in terms of land area, with over 8.5 million square kilometres. The wealthy cities of Brazil have always attracted migrants from the poorer regions of the country as well as from abroad. This is particularly true of Brazil's two largest cities, Sao Paulo (12.3 million) and Rio de Janeiro (6.7 million), both of which are located in south-east Brazil.

Brazil.

RURAL–URBAN MIGRATION

In developing world cities, such as those in Brazil, rural–urban migration continues at an alarming rate. People leave their rural homes and move to the city because of problems in the countryside (**push factors**) or because of the better opportunities they believe exist in the cities (**pull factors**). On arrival, migrants must find an unused piece of ground and build a shelter from whatever materials they are able to find. This leads to the creation of **shanty towns** which in Brazil are known as **favelas**. Initially, a favela may

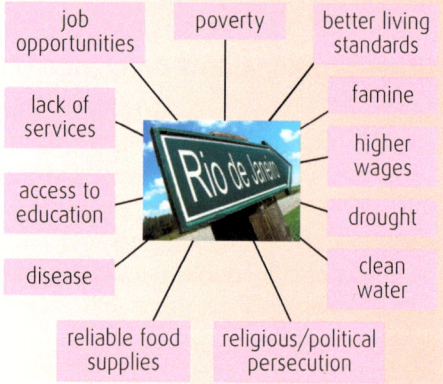
Diagram 2.18 Push and pull factors.

consist of nothing more than basic shacks lacking connections to sewage systems, water and electricity supplies. Usually, they are constructed on unwanted and unsuitable land which can be potentially dangerous, such as steep slopes, liable to landslides, or on river flood plains. However, once they have become more established, city authorities are often persuaded by residents to provide basic services such as water and electricity.

Favelas in Rio de Janeiro

It is estimated that there may be as many as 600 favelas in Rio de Janeiro, with 20–30% of the city's population living in them. In the 1970s and 1980s the city council evicted many favela residents, or **favelados,** and rehoused them elsewhere in the city before bulldozing the illegal settlements. However, this policy was highly unpopular and new favelas kept appearing as more migrants arrived in the city. More recently the city council, backed by the Brazilian government, has tried to improve living standards within individual favelas, by providing piped water supplies, sewage systems, rubbish collections and electricity. The favelas have been properly mapped with each alleyway being given a name, enabling residents to receive postal deliveries for the first time. Brazil's government-backed Growth Acceleration Programme is aimed at delivering improvements such as these.

Favelas in Rio and in other Brazilian cities have a bad reputation for crime, street gangs and in particular drug trafficking. Since 2004 the city authorities have established UPPs (*Unidade de Policia Pacifidora* or Police Pacification Units). Their purpose is to take back control of the favelas from the drug gangs. Occasionally, this has led to violence between police and organised criminal gangs, but an increasing number of favelas

Rocinha: A favela in Rio de Janeiro.

Rio de Janeiro's iconic skyline.

Human Environments – Urban geography: Cities in the developing world

have been reclaimed from the drug barons and are once again under the control of the authorities, with police regularly patrolling them. There was a particular push to reclaim favelas as safe areas before the city hosted the World Cup Final in 2014 and also the Olympic Games in 2016.

ROCINHA

Rocinha is the largest favela in Rio, and in the whole of Brazil, with an estimated population of over 70 000. It is also one of the longest established favelas and so the majority of properties now have basic sanitation, running water and also electricity. Many favelados have lived in Rocinha most of their lives and have made big improvements to their houses, giving it some of the best living conditions of all Rio's favelas. Situated within Rocinha are many businesses, banks, pharmacies and shops.

As an example of a favela which has significantly improved living conditions for residents, a comparatively new phenomenon has been the advent of guided tours into the heart of the favela. Rio de Janeiro is a tourist hot spot with millions visiting each year to see its world famous landmarks such as Copacabana beach, Sugar Loaf Mountain and the statue of Christ the Redeemer. Few dared to venture into the favelas before, but local tour companies today offer tourists the opportunity of taking a relatively safe guided walk into Rocinha. The favela tours have been seen as a good way of allowing visitors to see both sides of life in Rio, the rich and the poor, and also of showcasing some of the community volunteer projects that continually work to improve life for Rocinha's residents.

Complexo do Alemao

Complexo do Alemao is a group of favelas in the north of Rio de Janeiro. In common with most shanty towns, dwellings were built without any planning and there is no room for roads or vehicles. Rubbish has to be carried out of the favela. With properties so close together, there are few proper streets, only alleyways through which people have to walk, often for several kilometres to be able to get to other parts of the city, or just to get to work.

In 2011 a revolutionary transport system was opened, involving a system of cable cars. Some 152 gondolas, each with a capacity of ten people, run continually over the roof tops with six different stations. Residents are entitled to one free return ticket each day, although the current return price is less than £1. It runs for some 3.5 kilometres across the favelas and connects with the urban rail network in Rio. Its introduction revolutionised access for favelados, many of whom can now travel quickly to their jobs in other parts of the city. The project is part of Brazil's Growth Acceleration Programme, designed to improve the **infrastructure** of Brazil's cities.

Brazil is a land of contrasts, both in terms of its different landscapes and in the wealth of its people. Rocinha and Complexo do Alemao are just two examples of communities which are working hard to improve their own living standards.

DON'T FORGET

Shanty towns exist in many developing world cities. **Favelas** is the Brazilian name for them. Elsewhere in South America they are known as **barrios**, in South Africa as **squatter camps** and in India as **bustees**.

DON'T FORGET

A **favela** is a shanty town. A **favelado** is a person who lives in a favela.

Favela transport solution: cable cars.

Troops on standby in Complexo do Alemao – part of the Police Pacification programme of winning back favelas from gangs of organised criminals.

THINGS TO DO AND THINK ABOUT

1. Look at Diagram 2.18. Make a larger copy of the table opposite and place each **push** and each **pull** factor from the diagram into the correct column. There should be six in each column.

Push factors	Pull factors
Religious/political persecution	Better living standards

2. Using the information on these pages, make notes to describe how each of these initiatives can improve life in the favelas:
   - Police Pacification Units
   - Growth Acceleration Scheme
   - Cable Car system at Complexo do Alemao.

3. Do you think guided tours for visitors should operate in favelas such as Rocinha?

ONLINE

Check out the interactive urbanisation world map at www.brightredbooks.net/N5Geography

ONLINE TEST

Take the test on cities in the developing world online at www.brightredbooks.net/N5Geography

HUMAN ENVIRONMENTS

AGRICULTURAL CHANGE: FARM DIVERSIFICATION, GOVERNMENT POLICY AND GM CROPS

Farming uses approximately 70% of all the land in the United Kingdom, but employs less than 2% of the total workforce. Large areas of the UK, such as the Welsh mountains and Scottish Highlands, are unsuitable for arable farming but can still be put to agricultural use as pasture or rough grazing land for cattle or sheep.

The relatively small workforce is indicative of the fact that farming in Britain today is a highly efficient industry which makes use of the latest technology and needs comparatively few workers. Despite the large amount of imported food, British farmers still produce over 60% of all food consumed in the country. Due to globalisation and the free market within the European Union, farming is increasingly competitive. To stay competitive farmers have to be highly efficient, making the most of new technology and being prepared to venture into new activities and enterprises to maximise the farm's income.

DON'T FORGET

Diversification is not suited to every farm. Some farms have geographical advantages which increase their opportunities for diversification.

FARM DIVERSIFICATION

Farm diversification describes new ways in which farmers can earn income from their land, apart from their traditional farming activities. Often this might involve growing new crops or finding new outlets for farm produce, such as the opening of a farm shop to sell locally produced food or offering pick-your-own soft fruits, but often it involves branching out into non-farming activities. There are many examples of farm diversification – these include converting surplus property for new uses, such as holiday cottages or bunk barns; planting farm woodlands; offering services, such as riding lessons, pony trekking or archery; creating farm nature trails or adventure playgrounds; and opening up a field as a camping site or for use as car parking for nearby attractions. Some of the best known music festivals are held on land belonging to farms, for example, **Glastonbury** at **Worthy Farm** in Wiltshire, **T in the Park** at **Balado** in Fife and **Rock Ness** at **Clune Farm** on the banks of Loch Ness. It is estimated that over 50% of farms in the UK have diversified in some way.

Not every farm is suited to diversification. Much depends on the location of the farm and its proximity to population centres or to popular tourist areas. Larger farms usually have more opportunity for diversification and can afford to lose a small amount of land for new ventures. **Grant** funding from the government encourages many farmers to diversify and a Rural Development Programme in Scotland allows farmers to apply for financial assistance.

Factors affecting the suitability of a farm for diversification include:

- how close the farm is to major centres of population
- whether it is in an area with popular tourist attractions close by
- availability of possible grant funding from government
- accessibility of the farm
- the size of the farm and availability of land for new enterprises
- availability of labour

Farm shops and tea rooms.

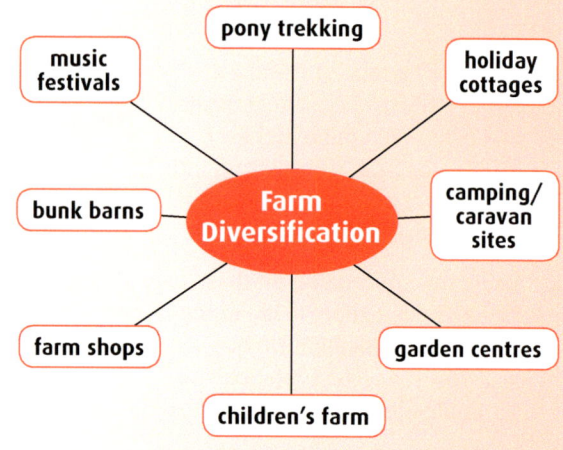

Diagram 2.19 Examples of farm diversification.

Human Environments – Agricultural change: Farm diversification, government policy and GM crops

GOVERNMENT POLICY

Farms are affected significantly by government agricultural policies. All European Union countries must follow the EU's **Common Agricultural Policy (CAP)**, which accounted for 36% of the total EU budget in 2019. The CAP aims to ensure a fair standard of living for farmers in return for a stable and safe food supply throughout the 27 countries of the EU. Originally, farmers were paid grants on the basis of how much they produced but this led to wasteful overproduction of some food supplies and so other criteria are now used. There are many different ways in which farmers can qualify for grants, but money is paid to them in the form of a single farm payment, based partly on the size of the farm. To get grants through the single farm payment, farmers must commit to respecting food safety standards, environmental protection, animal welfare and keeping the land in good condition.

80% of Scotland's land is agricultural, although much of it is hilly rough grazing land with poor soil and drainage. 85% of all farmland in Scotland is classed as belonging to part of a **Less Favoured Area**, entitling farmers to extra grants within their single farm payment. A Less Favoured Area is one that has significant geographical disadvantages such as poor weather, soils or a remote location.

Farms can also receive payments from Rural Development Funds which help farmers to diversify and set up new business ventures. Grants are available, for example, to help young farmers get started and for particular environmental schemes designed to help promote wildlife and benefit local ecosystems.

A less favoured area for farming: Rackwick, Hoy, Orkney.

GM CROPS

Genetically modified crops (GM) are crops that have been genetically engineered to make them grow faster, or in drier, cooler conditions for example. Their DNA is changed in a laboratory so that desirable traits, such as resistance to disease, can be built into the seeds. Although they are currently banned from commercial production in the UK and strictly controlled in the EU, they are well established world wide, especially in North America. In 2013 in the USA, 88% of all corn produced was GM and over 90% of all soyabeans. With the continual rise in global population, advocates of GM crops argue that without them, the world will not be able to feed itself by 2050.

GM crops continue to be a source of controversy with many people arguing that their widespread introduction is too risky for the environment. Some of the main arguments for and against GM crops are shown below.

Arguments in favour of GM crops include:

- higher crop yields
- can help to reduce food shortages, especially in the developing world
- fewer pesticides and herbicides needed
- improved food quality
- more resistant to weather extremes such as drought
- crops can be engineered to stay fresh for longer
- shorter growing time.

Arguments against GM crops include:

- seeds are more expensive to buy
- concern over, as yet unknown, adverse environmental effects
- biodiversity could be reduced; for example, by removing one crop pest, the food source for another animal might be unintentionally removed
- possibility of allergic reaction when consuming GM foods
- possible threat of new diseases emerging as some GM crops are modified by the use of bacteria and viruses.

DON'T FORGET

GM crops are not approved for commercial use in the UK and are very strictly controlled in the EU.

Oilseed rape is one of the main GM crops

ONLINE

Follow the link and check out the case study about how the Highland Council opposed GM trials at Roskill Farm on the Black Isle at www.brightredbooks.net/N5Geography

ONLINE TEST

Test yourself on these topics online at www.brightredbooks.net/N5Geography

THINGS TO DO AND THINK ABOUT

Make a list of different government policies which affect farming.

HUMAN ENVIRONMENTS

AGRICULTURAL CHANGE: ORGANIC FARMING

DON'T FORGET

Organic Farming is not an easy option. It involves a huge amount of work and can be unprofitable for the first few years.

ONLINE

Learn more about organic farming online at www.brightredbooks.net/N5Geography

ORGANIC FARMING: AN OVERVIEW

Organic farming is the production of food without the use of chemical fertilisers, herbicides and pesticides, genetically modified crops, or the routine use of drugs and antibiotics. Animal welfare is a cornerstone of organic farming as is wildlife conservation and caring for the environment. Although they are usually more expensive, organically produced food products have become increasingly popular as their perceived health benefits have become more widely known.

Organic farming has grown in the UK as a result of some of the worst excesses of **intensive farming**. Intensive farming is where the land is used to its maximum possible potential by the use of increasing amounts of chemical fertilisers and pesticides. In livestock farming this may include the regular use of antibiotics and drugs. Severe pollution of water courses has often resulted from chemical fertilisers being washed into rivers. Excess nitrates and phosphates can result in algal blooms which grow across the surface of ponds or lakes resulting in a lack of oxygen, killing fish and other wildlife. This process is known as **eutrophication** and harms the ecosystem. Animal diseases such as BSE, foot and mouth and bird flu have been attributed to some of the worst practices involved in intensive farming. It is partly as a result of this that the move towards organic food and organic farming has accelerated.

The aim of organic farming is to minimise any negative impacts that farming has on the environment and boost the positive ones. It aims to work with nature as much as possible. Whilst productivity is important, organic farmers make reducing the impact they have on the environment, when producing crops and livestock, a top priority.

The standards for organic farming are laid down by European Union law. All food products must meet these standards before they can be labelled as organic. One of the organisations which certifies products as organic in the UK is **The Soil Association**. The standards set for farms to qualify as organic are very detailed and strict. Some of these are as follows:

- The use of chemical fertilisers in the soil is banned, instead soil fertility should be maintained by growing and rotating a mixture of crops and adding organic matter such as compost or manure to the soil.

- The routine use of drugs and antibiotics on animals is banned.

- Laying hens must be free range at all times.

- Cattle should be allowed to graze outside naturally, weather permitting.

- The Soil Association requires that hedges must not be trimmed between 1st March and 31st August. This is to allow birds to nest.

To meet these standards, farms undergo an annual inspection. Good animal welfare and protection of the environment are key elements of this process. Also no genetically modified (GM) products can be used in organic farming.

To maintain soil fertility, farms cannot use artificial chemicals but instead must use organic fertilisers and **crop rotation**. This is where different crops are grown in each field in successive years. Crop rotation reduces the likelihood of crop diseases such as potato blight and helps to maintain a more healthy balance of minerals in the soil. In intensive farming, this balance is maintained by the addition of artificial fertilisers.

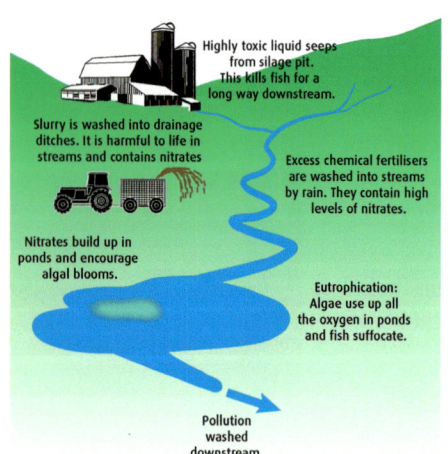

Diagram 2.20 How careless farming can harm the environment.

The Soil Association logo certifying a product as organic.

Human Environments – Agricultural change: Organic farming

ADVANTAGES AND DISADVANTAGES OF ORGANIC FARMING

A growing proportion of food sold in the UK is now labelled as organic. For the consumer **some** of the advantages of buying organic are:

- food products are free of the chemicals used in fertilisers, herbicides and pesticides
- organic foods often taste better
- the environment has not been harmed in their production, for example, bees and other insects are not affected by pesticides
- lack of chemical fertilisers and herbicides and the use of organic fertiliser (for example, manure) encourages the growth of soil microbes, making the soil healthier
- wildlife habitats have been protected and expanded
- it is more sustainable and reduces carbon dioxide emissions.

However, there are some disadvantages to the production of organic food and many people question whether it would be possible to feed the world's growing population of more than 7 billion people without intensive farming. The disadvantages of organic farming include:

- lower yields – without chemical fertilisers and pesticides productivity can be lower
- lower profits, at least to begin with
- organic farming can be more time consuming – weeds have to be physically pulled out or suppressed using a mulch (e.g. wood chips) between rows of plants
- a great deal of skill and patience is required to meet the standards for organic certification
- organic products are more expensive to buy.

In the UK, an average of just 3.5% of all agricultural land is classed as organic. This figure has fallen in recent years, reflecting the economic recession and expense and difficulty of establishing an organic farm. However, in some other European countries there is a higher percentage of organic farmland. Austria for example has 20% of its farmland classed as organic.

Yeo Valley Family Farm, situated in the Yeo Valley in Somerset, is the UK's largest organic farming business. It has a dairy herd of some 400 British Friesian cows whose milk helps to produce over 2000 tonnes of yoghurt each week as well as a variety of other dairy products. As well as being organic, Yeo Valley has diversified by running farm tours and opening up an organic garden tea room.

Yeo Valley HQ, by the Soil Association.

VIDEO LINK

Check out the video 'An organic farm in the Paris Basin' at www.brightredbooks.net/N5Geography

Yeo Valley family farm.

ONLINE TEST

How well have you learned about organic farming? Test yourself online at www.brightredbooks.net/N5Geography

ONLINE

Listen to the World Service programme 'Sustainable farming' to hear about the issues facing modern farmers at www.brightredbooks.net/N5Geography

THINGS TO DO AND THINK ABOUT

1. Explain why organic farming is better for the environment.
2. Find out more about organic farming and diversification by visiting these links:
   (a) Yeo Valley at www.yeovalley.co.uk
   (b) The Soil Association at www.soilassociation.org

HUMAN ENVIRONMENTS

AGRICULTURAL CHANGE: NEW TECHNOLOGY

DON'T FORGET

To remain competitive, farmers must apply for grants and make use of the latest technology where possible.

NEW TECHNOLOGY

Innovative new technology continually changes the way farms operate. The advent of mechanisation in the twentieth century saw a major revolution in farming with, for example, new tractors, combine harvesters and milking machines resulting in much more efficient working practices. This enabled farmers to complete work more quickly, cheaply and to increase their yields. The downside to these developments has been the loss of jobs in the farming sector. However, it is important for farmers to stay up to date with new technology in order to be able to stay competitive.

Polytunnels in the Wye Valley, used for growing soft fruits such as strawberries.

In the twenty-first century the development of new technology in farming is happening faster than ever. Often new technology might involve quite basic changes such as the use of polytunnels in fruit farming. The Carse of Gowrie in Tayside has acres of **polytunnels** producing hundreds of tonnes of soft fruit, such as strawberries and raspberries, each year. A polytunnel keeps the fruit warmer, allowing it to grow and ripen faster, while also being protected from ground frost in late spring. The market garden industry makes extensive use of polytunnels to produce salad vegetables. There are disadvantages to polytunnels including visual pollution and also the amount of plastic waste which is produced.

Precision farming makes use of the latest **GPS** (global positioning system) and **GIS** (geographical information system) technology together with scientific advances in soil testing. Soil in different parts of a field can be analysed and the minerals needed to achieve the right balance of nutrients in each area of a field indentified. Using highly accurate GPS devices, computers on board a tractor can then determine how much and what composition of chemical fertilisers should be added to each part of the field. This makes it possible for farmers to get the maximum possible **yield** from their land, increasing their profit margins.

Advantages of precision farming	Disadvantages of precision farming
Only the right amount of fertiliser is used. This reduces the cost to the farmer. Excess chemicals aren't washed into streams. Less harmful to the environment and more sustainable. Higher yields, bigger profits.	Cost of machinery and equipment is high. It may take several years for sufficient data to be collected to implement the system fully. Initially very labour intensive. Yields may not increase, but instead amount of fertiliser used may go down.

In the future it is possible that many farm activities will be controlled in this way, with computer-controlled tractors applying exactly the right amount of pesticide to individual plants according to which pest or bug the machinery identifies. Computer-controlled, driverless tractors using GPS already exist and could help to reduce labour costs, further increasing profit margins.

Hydroponics is the science of growing plants without soil. Instead, plants are cultivated indoors in a solution of water enriched with plant nutrients. Temperature and light can

contd

Human Environments – Agricultural change: New technology

be controlled giving the best possible conditions for growth. Hydroponics is already well established and is used to grow many vegetables now on sale in supermarkets. Using this system, growing conditions are strictly controlled, eliminating weather problems and enabling plants to grow very fast. There are fewer plant diseases and pests as everything is indoors. Vegetables grown this way even use less water than irrigated agriculture outdoors. **Aquaponics** is where fish farming techniques are combined with hydroponics, so for example, fish can be kept in the water used for growing plants. Fish such as tilapia produce nutrients through their waste and digestive system which plants then take up via their roots, eliminating the need for artificial fertilsers.

A further development of hydroponics is the concept of **vertical farming**.

Purpose built high-rise buildings use hydroponics to grow a variety of different plants to feed the local city population. It is estimated that a 30-storey high tower block could feed up to 50 000 people per year. Construction of a high-rise vertical farm is underway at Linkoping in Sweden, while smaller vertical farms, more like large greenhouses, already exist in the Netherlands and the USA.

Precision Farming: This display in a tractor cab allows the famer to see soil composition and crop yield data

- solves the problems of finding more land
- not all plants are suitable for growing using hydroponics
- there are many brownfield sites in cities
- very expensive to set up
- can be used to produce organic food
- reduces food miles as food is grown where it is consumed
- no insects, so pollination has to be done by hand: labour intensive
- less water used than in irrigated agriculture
- sustainable
- will take years to become profitable

Vertical farming: advantages & disadvantages.

 ONLINE TEST

How well have you learned about new technology in agriculture? Test yourself online at www.brightredbooks.net/N5Geography

 ONLINE

For more on vertical farming, check out the link at www.brightredbooks.net/N5Geography

 VIDEO LINK

Watch the talk on vertical farming online at www.brightredbooks.net/N5Geography

THINGS TO DO AND THINK ABOUT

1. Look at the photograph of the vertical farm and the information around it.

   Copy and complete the table below by placing each statement into the correct column. There are slightly more advantages than disadvantages.

Advantages of vertical farming	Disadvantages of vertical farming

2. Visit www.brightredbooks.net/N5Geography to find out more about hydroponics and vertical farming.

HUMAN ENVIRONMENTS

AGRICULTURAL CHANGE IN THE DEVELOPING WORLD 1

More than 80% of the global population lives in developing world countries and it is here that changes in farming techniques are likely to make the largest differences. Rapidly expanding populations put food supply under ever greater strain and so developments which can improve yields and productivity are vital.

The best agricultural developments are **sustainable** – they don't harm the environment or use up resources in a way that would disadvantage future generations.

GREEN REVOLUTION

During the second half of the twentieth century, a number of developments took place in developing world farming with the aim of improving food supplies for the rapidly expanding population. These developments were collectively known as the **Green Revolution** and included the introduction of **high-yielding varieties** of cereal crops (HYVs), new pesticides and chemical fertilisers, as well as new **irrigation** schemes and the production of cash crops. In response to concerns about the environmental impact of these developments, new sustainable, more environmentally friendly schemes have been called the **evergreen revolution**.

High yielding crop varieties included crops such as IR8 rice, developed initially in the Philippines to give much higher yields. It is claimed that without these crops food shortages and famine in the developing world would have been much greater. India's use of HYVs allowed it to change from a net rice importer to a leading world rice producer and exporter. However, large quantities of pesticides and fertilisers were necessary for the cultivation of HYVs and this had a negative impact on the environment. In particular, biodiversity was reduced as a result of the use of chemicals in rice paddy fields, resulting in fewer fish, frogs and other species. Human health was also put at risk as farmers were unaware of the risks of applying pesticides without protective clothing or chemicals polluting drinking water sources.

GM crop: golden rice compared with white rice.

HYV strains of rice and wheat were not originally genetically modified, but rather the result of careful seed selection and cross-breeding of different plant strains. More recently, true GM crops (where the plant DNA has been altered) have been introduced, with the amount planted increasing annually. GM now accounts for over 10% of all crops grown worldwide, including in many developing world countries such as Brazil, China, Pakistan and South Africa. India produces large amounts of GM cotton for example. Another GM crop which is

contd

Human Environments – Agricultural change in the developing world 1

currently being considered for approval in many developing world countries is **Golden Rice**. This is rice which has been enriched with beta-carotene to boost levels of vitamin A in areas where vitamin A deficiency is a serious health problem. Its producers claim there would be significant health benefits for the population in large areas of the developing world.

Irrigation in farming is where water is added to crops artificially instead of relying on natural rainfall. More widespread use of irrigation schemes has enabled crops, especially rice, to be grown in areas which were previously too dry. Small-scale local schemes are best as they are less expensive and can be controlled more easily by local farmers. The amount of land under agriculture has increased enormously as world population has risen, with India and China both having over half a million square kilometres of irrigated farmland. In some irrigation schemes water may drawn from beneath the ground, raising concerns about falling groundwater supplies and sustainability.

Other aspects of agricultural change in the developing world have included **land reform**, where, in Kerala state in India for example, land was removed from wealthy landowners and small amounts of land given to previously landless peasant farmers. Farmers have been educated in the use of modern farming techniques and encouraged to produce cash crops for extra income. In Kerala, this resulted in the production of crops such as coconuts, tapioca, ginger and other spices. In some areas there was so much emphasis on growing cash crops that there was insufficient food grown to feed the local population.

Irrigated rice terraces in Yunnan Province, China.

Diagram 2.21 Changes in Developing World Agriculture.

DON'T FORGET

The Green Revolution is the term used to describe major changes in developing world agriculture that took place in the second half of the 20th century. More small-scale changes, including appropriate technology and sustainable development taking place now, can be known as the evergreen revolution.

ONLINE

Read more about irrigated farmland in India at www.brightredbooks.net/N5Geography

ONLINE TEST

How well have you learned about agricultural change in the developing world? Test yourself online at www.brightredbooks.net/N5Geography

THINGS TO DO AND THINK ABOUT

1. Give the advantages and disadvantages of high yielding varieties of crops (HYVs).
2. Find out and write down all you can about one other change in developing world agriculture (see also pages 78–9).

HUMAN ENVIRONMENTS

AGRICULTURAL CHANGE IN THE DEVELOPING WORLD 2

ONLINE

Learn more about biofuels by reading the quick guide online at www.brightredbooks.net/N5Geography

BIOFUELS

Biofuels are produced using recently living organisms, such as plants or algae, to convert into energy. **Bioethanol** is an alcohol produced by fermentation of plants rich in carbohydrate such as wheat, maize, sugar beet and sugar cane. **Biodiesel** is made from crops which can produce suitable oils, such as oilseed rape, palm oil and soya, and is often made by mixing with recycled cooking oil. Biofuel production has expanded dramatically in the twenty first century, with producers promoting them as an alternative to dwindling stocks of **fossil fuels** such as oil, gas and coal. Brazil, China and India produce large amounts of biofuels.

A biodiesel powered bus.

Biofuel production can be controversial, particularly in developing world countries such as Brazil and Indonesia, where their cultivation has been associated with rainforest destruction. Opponents claim that not only has **deforestation** taken place in these countries but also that they don't help to reduce **greenhouse gas** emissions. Responsible biofuel producers claim that there is a significant reduction in emissions, depending on the method of production. A further problem has been that large areas of farmland have been used for growing biofuel crops, resulting in less food being produced for human consumption. This has led to rising world food prices and the possibility of global food shortages.

At a local scale, the collection of firewood by villagers in rural areas of developing world countries for use in cooking stoves is also seen as a cause of deforestation and **desertification**. As many as 2 billion people depend on this biomass as their main form of fuel. Not only can its collection be damaging to the environment but burning the firewood indoors causes serious respiratory problems. One solution in developing world countries such as India, has been to develop **biogas** – a fuel made from fermenting animal waste. Manure from cattle is collected, mixed with water in an underground tank and allowed to ferment. The methane given off in this process can be collected and used as gas for cooking, and the solid waste reused as fertiliser. Some three million households in India, Bangladesh, Pakistan and Nepal now have a biogas converter installed.

DON'T FORGET

Food shortages in the developing world and rising food prices in the developed world have been linked with the cultivation of biofuel crops. They may be grown on land which was previously used to feed people.

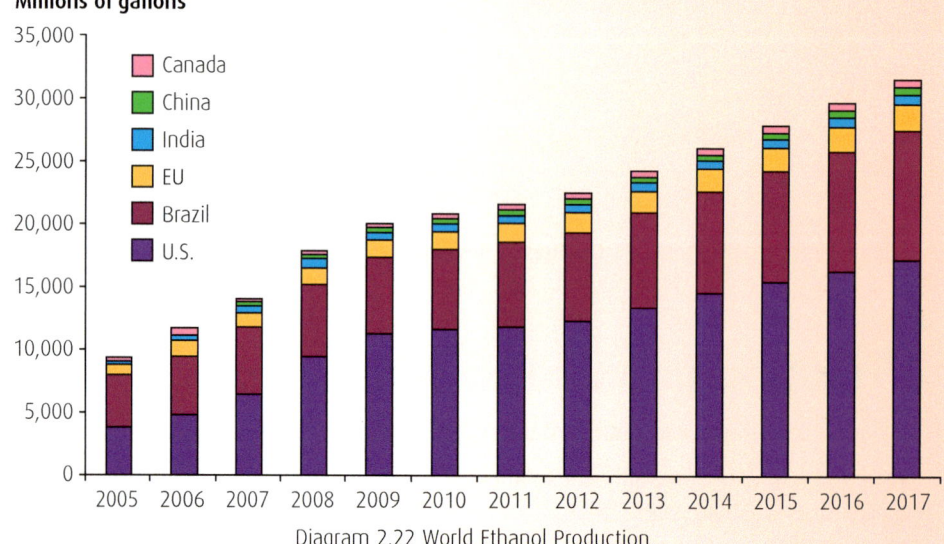
Diagram 2.22 World Ethanol Production.

Human Environments – Agricultural change in the developing world 2

APPROPRIATE TECHNOLOGY

Biogas is a form of **appropriate technology** – that is simple technology that works on a small scale and is not expensive. Appropriate technology is where sustainable development is promoted by developing projects which make use of local resources and knowledge and which are affordable. These small-scale projects often impact much more positively on developing world farmers than large-scale developments.

Many charities are involved in rural aid in developing world countries. One such charity is the British based organisation **Practical Action**.

This charity helps developing world farmers in numerous ways. One example of low-level technology used to help farmers in Bangladesh and India is by encouraging **rice–fish culture** (Diagram 2.23). Fish thrive in the dense foliage of the rice plants, safe from birds, while their droppings provide a source of organic fertiliser which helps the rice to grow. Farmers claim to have increased their rice yields by 10% through rice-fish culture, while they also have enough fish to provide regular high-protein meals for their families. Practical Action are involved in numerous new farming technologies (Diagram 2.24).

Biogas: appropriate technology.

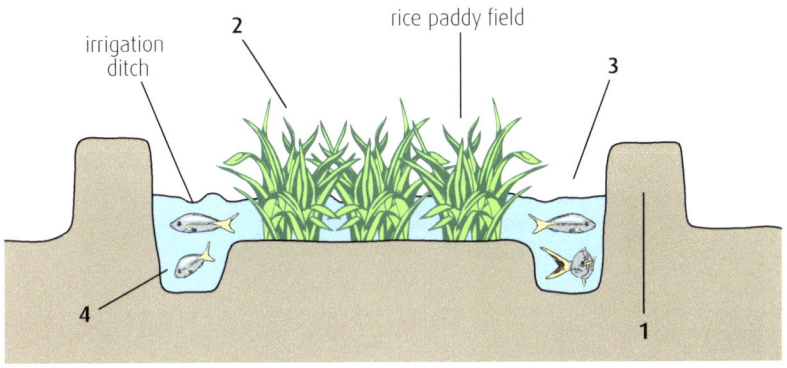

1. Earth walls (60cm high) built around field.
2. Rice seedlings transplanted into field with water about 12-15cm deep.
3. As rice starts to shoot, water level is raised and small fish are released.
4. Rice harvested, field drained and fish caught in ditches at field edges.

Diagram 2.23 Rice-fish culture.

ONLINE TEST

How well have you learned about agricultural change in the developing world? Test yourself online at www.brightredbooks.net/N5Geography

Diagram 2.24 Practical Action charity: examples of appropriate technology.

THINGS TO DO AND THINK ABOUT

1. For more information on appropriate technology and developing world farming, visit http://www.practicalaction.org
2. "Biofuels are good for the planet." Do you agree? Give as many reasons to support your answer as you can.

GLOBAL ISSUES

CLIMATE CHANGE: CAUSES AND EFFECTS

AN OVERVIEW

The Earth's climate has changed many times in the past. There have been numerous ice ages, the most recent one ending approximately 11 000 years ago.

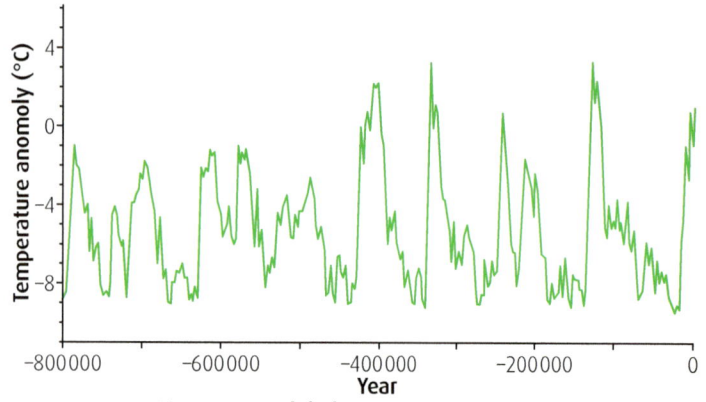

Diagram 3.1 Global temperature variation (data taken from ice core samples).

Evidence for this comes from ice core samples taken from, for example, ice sheets in Antarctica or Greenland. Contained within the ice are particles of windblown dust, seeds and tiny bubbles of air that can be analysed to show the composition of the atmosphere at the time they became trapped. From this information, the approximate global temperature and proportion of different gases which existed in the atmosphere at the time can be calculated. It has therefore been possible to construct a climatic record going back more than half a million years, showing that climate change is not new but has happened frequently (see Diagram 3.1).

Key terms

Greenhouse effect	Gases in the atmosphere trap heat
Global warming	Increasing average global temperatures
Climate change	Alterations to average global weather patterns

EVIDENCE OF CLIMATE CHANGE

There is little dispute that global temperatures rose by an average of about 0.8°C in the last century. This could be part of the normal cycle of temperature change between ice ages. Much more significant is the level of carbon dioxide (CO_2) increase. CO_2 is a **greenhouse gas** that traps heat in the atmosphere. In 2021 CO_2 levels reached 417 parts per million, a figure significantly higher than at any time measured in the past (see Diagram 3.2). Together with the current trend for rising temperatures, this is compelling evidence of climate change. Other greenhouse gases include methane, nitrous oxide, ozone and chlorofluorocarbons (CFCs). Increases in the proportion of these gases in the atmosphere would be further evidence of climate change; methane levels, for example, have increased significantly in recent decades.

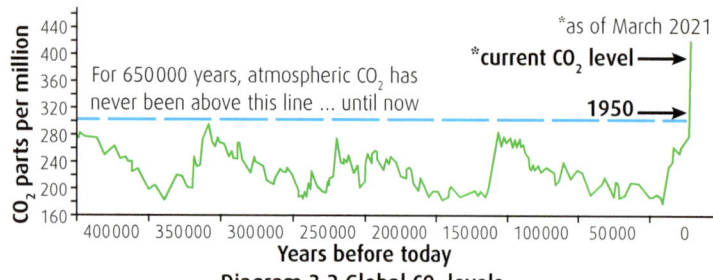

Diagram 3.2 Global CO_2 levels.

Other evidence of climate change includes a 17 centimetre rise in global sea levels during the last century, warming surface temperatures in the world's oceans, shrinking ice sheets in Greenland and Antarctica, a dramatic decrease in the thickness and extent of Arctic sea ice in recent decades, retreating glaciers in all of the world's mountain ranges and an increase in the frequency of extreme weather events.

DON'T FORGET

The greenhouse effect is a natural process in which a blanket of greenhouse gases in the atmosphere traps heat, making the Earth warm enough for us to live on. By adding extra gases to the atmosphere, humans have increased the greenhouse effect by making the blanket of gases thicker, causing global warming.

PHYSICAL CAUSES

Today, the term **climate change** is often used to describe alterations to the Earth's atmosphere for which humans have been responsible. Clearly, because there have been a number of ice ages, climate change has happened in the past, long before the size of human population was an issue. There are therefore many possible physical causes of climate change as well as those which might be caused by humans.

Physical causes of climate change can be attributed to alterations in the global pattern of ocean currents, volcanic activity, variations in the amount of solar radiation as well as changes in the Earth's orbit around the Sun. **Milankovitch theory** describes several

contd

Global Issues – Climate change: Causes and effects

possible changes to the Earth's orbit that would result in variations to solar radiation received on the planet. For example, over a period of 42000 years the angle at which the Earth's axis tilts changes between 22.1° and 24.5° and back again (it is currently 23.5° and decreasing), while the orbit of the planet around the Sun varies between an elliptical and a circular shape over a period of 96000 years. Clouds of ash, dust and gas emitted by volcanic eruptions can result in the amount of solar radiation reaching the Earth's surface being reduced, while extra greenhouse gases, such as CO_2 and methane, can lead to **global warming**. The eruption of Mount Pinatubo in the Philippines in 1991 emitted huge amounts of dust into the atmosphere, resulting in a 0.5°C drop in global temperature in 1992.

HUMAN CAUSES

Ninety-seven per cent of academic research carried out in the last 20 years agrees that human activity is responsible for climate change. Burning **fossil fuels** in power stations, factories and transport emits billions of tonnes of CO_2 every year. **Deforestation** releases CO_2 when forests are burned and reduces the amount of vegetation that can absorb CO_2 from the atmosphere. Clearing land for agricultural use can release CO_2 from the ground. Some 30 billion tonnes of CO_2 are released into the atmosphere by human activities each year.

Agriculture is also the source of large amounts of methane, emitted from rice paddy fields and raising livestock such as cattle. The impact of methane as a greenhouse gas up to 30 times greater than that of CO_2, although it doesn't last as long in the atmosphere. Methane also comes from domestic rubbish in landfill sites and from leaks in the extraction of oil and natural gas.

Nitrous oxide is emitted when fuel is burned and also from agricultural activities. CFCs are produced in industry and used in many areas such as in coolants, fertilisers and as foaming agents in fire extinguishers. The increase in the human population and consequent rise in agricultural activity is certainly a major cause of climate change.

EFFECTS OF CLIMATE CHANGE

If average temperatures rise as predicted, by between 2 and 5°C, the overall effects on the planet are highly likely to be catastrophic. Effects will include further loss of sea ice, a more rapidly rising sea level and more frequent and intense weather extremes such as heat waves and wild fires. Although there may be temporary benefits as the climate changes in some locations, such as the possibility of growing new crops, the effects of climate change are likely to be disastrous in most places.

The most devastating effects are likely to be in developing world countries, where drought and coastal flooding will continue to affect millions of people. However, the developed world will also be further affected by coastal flooding. Water shortages will become more severe as groundwater supplies fall and glacial retreat in areas such as the Himalayas will result in a water deficit for large parts of the Indian subcontinent. The effects on agriculture could be equally serious as droughts and water shortages could reduce yields by up to 50% in some areas of Africa in this decade.

There is already evidence of a reduction in biodiversity and the extinction of many animal species. The shrinking Arctic ice cap is threatening the survival of polar bears, who depend

DON'T FORGET

Your **carbon footprint** is the impact that you have on the atmosphere through the production of greenhouse gases associated with your lifestyle.

Effects of climate change: Franz Josef Glacier, New Zealand in 2009 (left) and 2013 (right).

DON'T FORGET

World sea levels are rising because as water close to the ocean surface warms up, it expands. This is **thermal expansion** and is likely to be the main reason for coastal flooding. Melting ice from Antarctica and Greenland will also contribute.

Effects of climate change: Kivalina, Alaska. The Inuit settlement of Kivalina: set to disappear by 2025 as sea erosion threatens to wash it away now that the sea ice is not there to protect it in storms.

ONLINE

For more evidence of climate change and the effects of global warming go to www.brightredbooks.net/N5Geography

ONLINE TEST

Test yourself on the causes and effects of climate change at www.brightredbooks.net/N5Geography

VIDEO LINK

For more on climate change, check out the video clip at www.brightredbooks.net/N5Geography

THINGS TO DO AND THINK ABOUT

1. Copy and complete this table by adding as many physical and human causes of climate change as you can.

Physical causes of climate change	Human causes of climate change

2. Make a star diagram to show the effects of climate change.

GLOBAL ISSUES

CLIMATE CHANGE: MANAGEMENT

DON'T FORGET

At the Paris Climate Change Conference in 2015, some 195 countries agreed to do everything possible to limit global warming to no more than 2°C above pre-industrial levels.

DON'T FORGET

Other forms of renewable energy such as hydro-electric, wind, geothermal, wave and tidal energy are key in helping to manage global warming by reducing carbon emissions.

ONLINE

Calculate your own carbon footprint at www.brightredbooks.net/N5Geography

ONLINE TEST

Test your knowledge of this topic at www.brightredbooks.net/N5Geography

MANAGING CLIMATE CHANGE

Environmental groups have been raising awareness about the threat posed by climate change for decades. Many strategies are aimed at reducing energy consumption in order to minimise emissions from fossil fuels. On a global scale there have been a number of international conferences organised by the United Nations (UN) to highlight the problems of climate change and the need for action. Some of the most significant meetings, to which almost every country on the planet sent representatives, have been the:

- **UN Earth Summit**, Rio de Janeiro in 1992
- **UN Climate Change Conference**, Doha, Qatar, 2012
- **UN Climate Change Conference**, Paris, December 2015
- **UN Climate Change Conference**, Glasgow (COP 26), November 2021

The Paris Climate Change Conference at the end of 2015 marked a major milestone in global cooperation to tackle the causes of climate change. Virtually every nation in the world (196 countries) agreed to try to limit average global temperature increases to 1.5°C of pre-industrial levels by 2100. This will be achieved by reducing greenhouse gas emissions around the world, and the use of renewable energy sources to replace fossil fuels. Many countries will need to achieve zero emissions sometime between 2030 and 2050 in order to meet the target of 1.5°C. After initially backing out of it, the USA has rejoined and again committed itself to the goals of the agreement in 2021.

As large government grants become available for developing carbon neutral power supplies the alternative energy sector is booming. Investment in alternative energy is also seen as important in providing fuel security for countries amid concern that fossil fuels will inevitably run out. At a local level, councils are encouraging environmentally friendly transport initiatives and recycling schemes aimed at reducing waste and unnecessary consumption. Individuals are encouraged to be more environmentally responsible in their everyday lifestyles by making simple choices such as walking or using public transport for short journeys, shopping with reusable instead of plastic bags and switching off lights and other electrical appliances that are not needed. All of these initiatives support the idea of **sustainable development** and cutting back on global greenhouse gas emissions.

REDUCING GREENHOUSE GAS EMISSIONS

There is international acceptance that governments should do as much as they can to reduce greenhouse gas emissions. The Reuse, Reduce, Recycle logo (Diagram 3.3) reflects the need for sustainable environmental policies, which increasingly affect every aspect of life. There are numerous different strategies that have been adopted in many countries to meet these aims. They include schemes to:

- prevent deforestation
- plant more trees (**afforestation**)
- develop alternative fuels for cars
- promote renewable energy (see pp40-41)
- develop green transport policies
- improve house insulation
- increase the consumption of locally produced food to reduce food miles.

Diagram 3.3 The Reuse, Reduce, Recycle logo.

Developing clean alternative energy sources is key to reducing greenhouse gas emissions. Some of these are described on pages 40 to 41 Examples of other methods of reducing greenhouse gas emissions are described on page 83.

Home insulation and heating

In countries with cold winters, fully insulating houses can make a huge difference to the amount of energy needed to keep properties warm. In the UK, the reduction of greenhouse gas emissions has become a priority. For example, government and local council grants of up to 100% have been available for improving loft insulation,

contd

Global Issues – Climate change: Management

fitting double glazing and the installation of solar panels. New ways of heating houses, including ground source heating, have been developed.

Ground source heating makes use of the warmer temperatures which exist underground in winter. By circulating water through underground pipes and using a pump to bring the heated water back to the surface, it can be used to heat the property. Ground source heating is expensive to install and needs an external power source to operate the pump, but it still greatly reduces the need for other fuel sources.

Charanka Solar Power Station, Gujurat, India.

Solar panels have been used for many years in warm countries, but with their increased efficiency they are becoming more common in less sunny climates such as the UK. Solar panels can be used to provide not only electricity but also hot water. Solar power stations are being built around the world in countries such as Spain, the USA, China and India. The Charanka Solar Power Station in Gujurat, India, for example, will have a capacity of 600 MW on completion (see photo). Solar power is likely to become a major contributor to reducing carbon emissions.

DON'T FORGET

Biofuels have also been developed to reduce harmful emissions from vehicles.

Sustainable transport

In the UK 27% of all carbon emissions come from transport. The rapid growth in the ownership of private cars has led to big increases in greenhouse gas emissions. Industry requires diesel-hungry heavy goods vehicles for much of its transport requirements, while head for head, air travel is responsible for more greenhouse gas emissions than any other form of transport.

DON'T FORGET

Electric cars are only environmentally friendly if the electricity used to recharge the battery is generated without emitting greenhouse gases.

Attempts to reduce these emissions have revolved around developing much more efficient engines and encouraging people to travel less or in more environmentally friendly ways. New airliners such as the Boeing 787 Dreamliner and the Airbus A350 have been built partly from composite materials to reduce weight and improve fuel efficiency. Car companies are increasingly launching all-electric or hybrid vehicles, including some running on alternative fuels such as hydrogen (see photo). Cities are increasingly adopting policies aimed at promoting more environmentally friendly modes of travel. Encouraging people to use public transport and leave their cars at home is key to this.

Locally produced food

One impact of globalisation has been the transport of foodstuffs over thousands of miles from producers to consumers. The term **food miles** has been coined to describe this and to highlight the amount of energy used (and greenhouse gas emissions) to transport and store food until consumption. Producers encourage people to buy locally as it is perceived as being good for the environment due to a lower carbon footprint (as well as being good for the local economy). Many people would agree that it makes sense to buy locally produced food where possible.

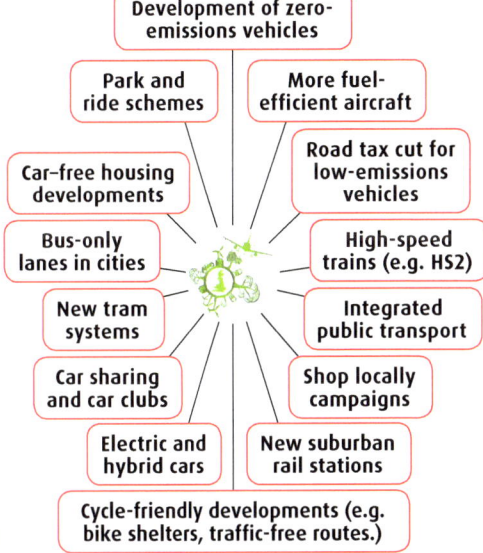
Diagram 3.4 Developments aimed at reducing greenhouse gas emissions from transport.

However, food transported over a long distance is not necessarily less environmentally friendly than locally produced food. Tomatoes grown naturally in Spain and transported to the UK, for example, may have a smaller carbon footprint than UK tomatoes grown in heated greenhouses. It is important to consider the method of production as well as the distance over which goods have been transported in order to gain a fair comparison. Exports of food (cash crops) from many developing countries may also benefit some of the world's poorest communities, particularly if it is fair trade produce. Buying locally produced food may help to reduce carbon emissions but there may be other considerations to take into account.

Hydrogen-powered London bus.

 VIDEO LINK

Check out the video 'Potatoes from Egypt' at www.brightredbooks.net/N5Geography for an insight into food miles.

 THINGS TO DO AND THINK ABOUT

1. What things can you do at a **personal** or **local** level to help reduce greenhouse gas emissions?
2. How can a government help the whole country to reduce greenhouse gas emissions?

GLOBAL ISSUES

TUNDRA REGIONS

The term **tundra** comes from the Finnish word 'tunturia' which means treeless desert. Key characteristics of tundra regions are:

- cold climate
- low precipitation
- permafrost with waterlogged top soil in summer
- very short growing season
- no trees, low-growing plants
- few species of plants and animals
- very sparsely populated.

DON'T FORGET

The **temperature range** is the difference between the average maximum and average minimum temperatures. Tundra climate has a large annual temperature range.

CLIMATE AND LOCATION

Tundra regions are areas of cold, dry climate found mostly in the Arctic. Average temperatures remain below zero for eight or nine months of the year, with summer averages rarely exceeding 10°C. Snow lies on the ground for most of the year, but total amounts of precipitation are low and tundra areas are sometimes referred to as cold desert. In summer, there are several weeks when the Sun never sets, but is low in the sky and its rays are comparatively weak. In winter, there are a corresponding number of weeks of total darkness when the temperatures can fall to extreme lows.

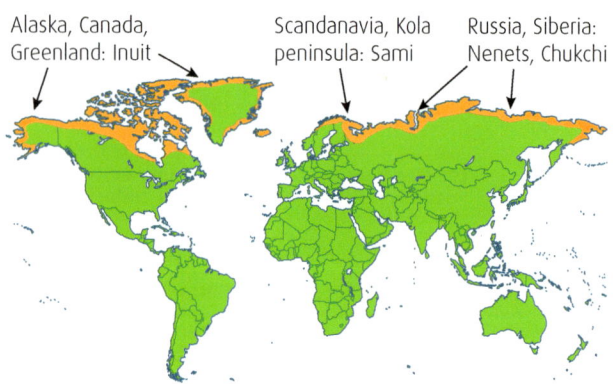

Diagram 3.5 Map of tundra regions and indigenous people.

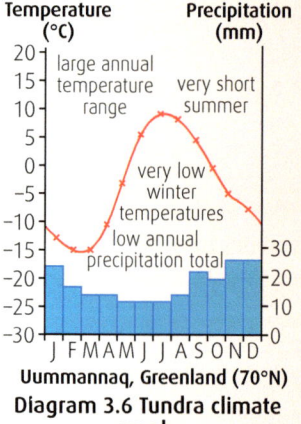

Diagram 3.6 Tundra climate graph.

The main tundra areas are found in northern Alaska and Canada, the coast of Greenland along the edges of the ice cap, parts of northern Scandinavia and along the northern coasts of Russia and Siberia. Elsewhere, tundra conditions can be found in mountainous regions where the climate is cooler due to altitude such as on the Cairngorm plateau.

THE TUNDRA ECOSYSTEM

Diagram 3.7 Tundra vegetation adaptations.

- Low growing to escape high winds and colder temperatures
- Short roots to avoid permafrost
- Can grow in thin acidic, rocky soil
- Can extract minerals from bare rock
- Germinate, flower and seed very quickly
- Snow cover provides insulation through long winters
- Small leaves, dense, compact foliage help conserve warmth

Tundra regions are cold, barren and windswept, with a growing season that lasts for three to four months at best. There is a deep layer of permanently frozen soil or **permafrost** that prevents plant growth for most of the year. Only the top few centimetres melt in summer, creating a boggy landscape of marshes, ponds and low-growing shrubs and plants. In northern Canada and Alaska, this waterlogged soil is known as **muskeg**. Water cannot drain down through the soil because of the permafrost and so it lies on the surface.

Tundra lands are treeless as the permafrost prevents their roots from developing and frequent high winds stunt their growth. Vegetation is adapted to this harsh climate by germinating and completing its annual life cycle very quickly during the short summer season. Plants have to be low growing and tolerant of waterlogged conditions as they grow in poor acidic and rocky soil. In summer, plants and small shrubs

contd

such as sedges, cotton grass, lichens, mosses, crow- and bearberries flower briefly and dramatically before the temperatures fall permanently below zero again. Tundra plants are also adapted to the cold temperatures by having small leaves and they depend on the snow cover in winter to keep them insulated from the intense cold.

The tundra ecosystem is fragile and vulnerable to disturbance by people or climate change. There are swarms of insects, such as mosquitoes, in summer and migratory birds come to feast on them. Geese are found in large numbers during the summer months and the snowy owl is resident in many tundra areas. There are relatively few mammals which can survive the harsh tundra conditions but caribou (North America) and reindeer (Europe) are native to tundra regions, with vast herds migrating further northwards in summer to feed on the mosses and lichens. Polar bears, Arctic foxes, wolves, musk oxen, wolverines and lemmings all have very thick coats to provide insulation against the intense Arctic cold.

The musk ox, for example, has very fine short hair close to its body that is warmed by its own body heat, and a long outer layer of fur that insulates it from wind and water. The fur of many resident animals, such as the ptarmigan, Arctic hare, Arctic fox and Arctic wolf, changes to white for camouflage through the long winter months. Many creatures, such as the ptarmigan, will burrow down into the snow and use it to insulate them against the intense cold, while others, such as the Arctic hare and Arctic wolf, have wide feet and long legs to keep them from sinking into the snow.

All tundra animals have to raise their young quickly during the short summer season. Animals feed intensely through the summer while food is more abundant to build up fat reserves in preparation for winter. Reindeer may have up to 30% extra body fat by the end of the summer which will help them survive long spells when there may be little or no food available in winter.

Diagram 3.8 Reindeer are adapted to survive in the tundra.

INDIGENOUS PEOPLES

Tundra regions are very sparsely populated by people whose traditional life styles have changed in recent times as a result of persecution and loss of their ancestral lands.

The **Inuit** live in the tundra and high Arctic areas of Alaska, Canada and Greenland. The **Sami** inhabit northern Norway, Sweden and Finland and the Kola Peninsula of Russia, while the **Nenets** and **Chukchi** people are found in parts of northern Russia and Siberia.

In Greenland the Inuit have traditionally survived by hunting. They depend on fishing, ice fishing and hunting for mammals such as seals. They are not able to grow anything due to the low temperatures and short growing season. In summer, hunting and fishing are carried out in **kayaks**, while in winter dog sledges are used to venture out on to the sea ice to hunt for seals, where they surface for air through holes in the ice. Inuit clothing originally consisted entirely of animal skins which were skillfully sewn to keep them warm and dry in extreme weather. Inuit life is characterised by hardship, with long and dangerous journeys to find food, often in extreme and life-threatening weather conditions. Expertise and knowledge handed down from generation to generation helps the Inuit to navigate and find enough food in the most challenging of environments.

 ONLINE

To find out more about how reindeer are adapted to tundra conditions go to www.brightredbooks.net/N5Geography

 ONLINE TEST

Take the tundra regions test at www.brightredbooks.net/N5Geography

 VIDEO LINK

For more on how climate change is affecting the Arctic, watch the clip 'Evidence for global warming – polar ice caps' at www.brightredbooks.net/N5Geography

THINGS TO DO AND THINK ABOUT

1. Describe tundra climate. You should refer to seasons, maximum and minimum precipitation, temperatures and temperature range.
2. Draw a star diagram to show how the Inuit have adapted to living in the tundra.

GLOBAL ISSUES

TUNDRA: EFFECTS OF HUMAN ACTIVITY

The tundra lands are among the most remote and sparsely populated areas on Earth. Their hostile climate, inaccessibility and lack of daylight in winter make settlement difficult. Over many generations, **indigenous people** such as the Inuit, Sami and Nenets have built up extensive knowledge of how to survive in the tundra. They appreciate the fragility of the environment and know that even the slightest changes can have a big impact on the ecosystem.

Increasingly, the tundra is under threat from human activity on both a global scale and a local scale. The initial effects of climate change are already evident in many parts of the Arctic, where average temperatures are rising faster than anywhere else on the planet. Rich mineral deposits on land and off the coasts of tundra areas attract industrialists keen to exploit them. Melting ice caps and permafrost make mining and drilling easier and more profitable. Tundra environments are under threat from land degradation more than ever before.

DON'T FORGET

Snow-covered sea ice reflects 90% of the Sun's radiation, whereas the open ocean absorbs 90% of solar radiation.

CLIMATE CHANGE

The Arctic has experienced some of the largest temperature increases due to global warming so far, and predicted increases by 2100 could be higher than 5°C above 1990 averages. As average temperatures rise, the amount of snow cover in the tundra is reduced, exposing rock, soil and vegetation, all of which are darker in colour than snow and more likely to absorb heat. As the permafrost melts, CO_2 and methane trapped in the frozen soil are released into the atmosphere. Methane is 30 times more powerful as a greenhouse gas than CO_2, so this is a matter of real concern. Arctic sea ice has been melting back at an unprecedented rate. Since accurate measurements have been possible due to the advent of satellite technology from about 1979, the extent of Arctic sea ice has decreased by an average of 3% per year. In September 2012 there was a new record low recorded.

As reflective Arctic sea ice melts back exposing the dark surface of the ocean, much more radiation from the Sun is absorbed, further accelerating the likely rate of global warming. It is predicted that the Arctic Ocean could be ice free in summer before 2030.

Sea ice reflects the Sun's rays, open water absorbs them.

The continued melting of sea ice will affect marine and tundra ecosystems as animals, such as polar bears, seals and whales, have fewer places to hunt and to rest. Polar bear populations are predicted to fall by up to two thirds if Arctic sea ice melts back as forecast. Coastal communities in the tundra are affected by rising sea levels and increased erosion. Traditional hunting communities are no longer able to venture out on to the sea ice, which is too thin and forms for less time each year.

ONLINE TEST

Test yourself on the effects of human activity in the tundra region online at www.brightredbooks.net/N5Geography

EFFECTS OF HUMAN ACTIVITY IN THE TUNDRA

Scientific exploration is a common way in which people use the tundra. Research is carried out into climate change and the likely effect of continued warming. By studying ice cores and permafrost, scientists can predict the likely impact of global warming on the tundra ecosystem. Ecologists study how plants and animals are adapted to tundra conditions and how human-induced climate change might affect them.

VIDEO LINK

Check out the video 'Observations on Climate Change in the Arctic – WWF' at www.brightredbooks.net/N5Geography

Tourism in tundra areas is increasing as visitors seek to experience the Arctic wilderness for themselves. Fishing, hiking and kayaking tours are popular in summer, and in winter skiing and dog sledging are in demand. Game hunting has caused a decline in musk ox and caribou numbers in Canada and Alaska, for example, and is now strictly controlled by a licensing system. Subsistence hunting by indigenous groups never posed the same threat to the tundra ecosystem. Cruise liners skirt the coastlines of tundra areas, such as the Lofoten Islands of northern Norway, allowing passengers to view the stunning scenery, the wildlife and see the night time displays of the aurora borealis. In many areas,

contd

Global Issues – Tundra: Effects of human activity

national parks and nature reserves have been established to protect the fragile tundra environment. Tourism can bring an economic boost to remote tundra settlements but can also pose a threat as new roads, hotels and water and waste disposal systems are developed to cope with increasing tourist numbers.

Mining, drilling and exploration for minerals, in particular for oil and gas, are perhaps the biggest threat to tundra regions and their communities after climate change. This sort of activity can be very damaging to delicate tundra ecosystems as oil wells are sunk, test boreholes drilled and the ground excavated to reach deposits of uranium ore, for example. New roads, pipelines and buildings drive away wildlife and can disrupt the migratory patterns of animals such as caribou. Vehicles compact the soil and destroy the fragile vegetation. Pollution from oil leaks on land and oil spills at sea have been immensely damaging. Air and water pollution caused by air-borne pollutants such as PCBs and insecticides blown in from elsewhere build up in the ecosystem and become poisonous to plants and insects. Levels of these pollutants increase further up the food chain and can have a damaging effect on wildlife. Lichen are particularly sensitive to air pollution and their growth can be adversely affected. Any reduction in the availability of lichen affects the entire tundra food chain.

Diagram 3.9 Reindeer being herded to new grazing land across a sea inlet in northern Norway.

OIL POLLUTION IN THE KOMI REGION OF RUSSIA

Northern Russia and Siberia are rich in mineral resources including precious metals, uranium ore and fossil fuels. Oil exploration and drilling have been carried out in the Komi region of northern Russia for decades. In the area around Ust-usa on the Pechora river and its tributary the Kolva, there are over 3000 drill holes, thousands of kilometres of pipelines and service roads built by the oil industry. Unfortunately, there is a poor record of oil spills and leakages resulting in environmental degradation. Often, oil spills go unreported and leaks from ageing pipelines continue to ooze oil into the tundra soil.

Leaking oil forms toxic lakes, destroys the vegetation and seeps into rivers and groundwater. Permafrost prevents oil from seeping into the ground so it spreads further and low temperatures stop it from evaporating. A further hazard is the extensive flaring of oil and gas, which not only releases CO_2 into the atmosphere, but also leaks deadly toxins into the surrounding environment.

The local Komi people, whose way of life includes hunting, fishing, gathering berries and reindeer herding, have found their existence threatened. Fish have been poisoned in rivers and reindeer will not eat moss and lichens contaminated by oil. Reindeer herds have to be taken further and further afield to find clean grazing lands. Another concern is that leaking oil is washed by the Pechora river into the Arctic Ocean, causing serious marine pollution. Remote Komi communities are threatened by extensive environmental pollution as their way of life becomes increasingly unsustainable.

Globally, there is real concern about the possible expansion of oil exploration into areas of the Arctic Ocean that are no longer covered by ice. The Exxon Valdez (Alaska, 1989) and Deepwater Horizon (Gulf of Mexico, 2011) disasters are examples of oil-based catastrophes and, fearing the possibility of such an event in the Arctic Ocean, environmental campaigners are calling for a ban on further **exploitation** of oil reserves there.

 ONLINE

Read more about oil pollution in the Kolva River in Russia at www.brightredbooks.net/N5Geography

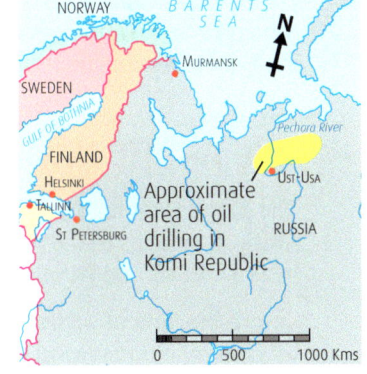

THINGS TO DO AND THINK ABOUT

1. List ten ways in which tundra landscapes can be used. Highlight those uses which do not pose a threat to the tundra environment.

2. Make a star diagram to show how oil pollution can affect the environment of the tundra and the livelihood of indigenous people.

GLOBAL ISSUES

MANAGING TUNDRA ENVIRONMENTS

Large areas of the tundra are under threat from different human-induced pressures. These include:
- climate change
- exploitation of mineral resources including fossil fuels
- tourism.

PROTECTING TUNDRA ENVIRONMENTS

As tundra regions are so remote and inaccessible, they receive little media attention and often environmental damage may go totally unreported or, at best, get minimal news coverage.

At a world level, the UN has been important in attempting to get international agreements on limiting greenhouse gas emissions to slow down climate change (e.g. the Kyoto Protocol in 1997) but this is notoriously difficult and time-consuming. However, the UN has also set up a World Network of Biosphere Reserves, some of which are in tundra areas (e.g. the Taimyrsky Biosphere Reserve in northern Russia). This has helped to identify areas at risk and to give them a greater chance of protection.

National governments also play a role in protecting the environment by introducing laws designed to stop developments that will pollute tundra areas. However, these laws are often poorly enforced and sometimes have little impact on stopping development from damaging the environment. A more successful area of government intervention has been in the setting up of **national parks** and **nature reserves**, where very strict planning regulations prevent or carefully control any new developments in the tundra. There are many examples of protected areas in the tundra such as the Ivvavik and Vuntut National Parks in Canada, the Urho Kekkonen National Park in Finland and the Arctic National Wildlife Refuge in Alaska, USA.

It is through the work of scientific research carried out by some of the world's leading universities and the efforts of environmental organisations, such as **Greenpeace**, the World Wide Fund for Nature (WWF) and the Earthwatch Institute, that many of the environmental issues affecting the tundra today are brought to public attention. These are non-governmental organisations (not run by the government) and non-profit making charities. They depend on charitable donations from the public to be able to carry out research and operate campaigns to bring environmental problems to the attention of the public. Many changes have been brought about by their campaigns. For example, in 2013, the government in Greenland agreed not to issue any new licences for offshore oil exploration around its coasts.

THE ARCTIC NATIONAL WILDLIFE REFUGE

The Arctic National Wildlife Refuge (ANWR) in Alaska, USA, is the largest area of protected land anywhere in the Arctic. It belongs to the people of the USA and was first established in 1960 and then enlarged in the 1980s. It is run by the US Fish and Wildlife Service and its main aims include:

- preservation of wilderness areas
- conservation of plants and animals in their natural environment
- allowing indigenous people to continue their hunting and gathering activities
- protecting water quality and resources.

In total, it covers over 78000 square kilometres, which is about the same size as the whole of Scotland. Much of this land contains tundra, with just two isolated settlements at Kakotovik and Arctic Village belonging to the indigenous **Inupiat** and **Gwich'in** peoples, respectively. There are no roads, marked trails or camping places. About 72000 square kilometres of this land is classed either as wilderness area or an area of minimal management. These areas are protected by law from all development and must remain natural. The refuge includes part of the Brooks Mountain range and two so-called wild rivers. There is also a coastal plain which adjoins the Barents Sea and extends to some 6000 kilometres. At present, there is a debate in the USA about whether exploration and drilling for oil and gas should be allowed in this area. Just to the west of the ANWR at Prudhoe Bay is one of the most productive oilfields in the USA, supplying more than 25% of its oil, and geological evidence suggests that there are also likely to be large oil reserves in the coastal plain of the ANWR.

The ANWR is a vast, wild and treasured landscape. Ecological processes and natural diversity are sustained, making it a living laboratory where the natural Arctic and subarctic environment can be studied and appreciated.

One of the main goals of the ANWR is to be able to pass on the refuge undiminished to future generations. Diagram 3.10 shows some of the ways in which the refuge is managed in order to be able to achieve this goal.

Global Issues – Managing tundra environments

Caribou migrating across the tundra of the ANWR

Diagram 3.9 Map of the Arctic National Wildlife Refuge.

Diagram 3.10 ANWR: protecting the tundra environment.

IVVAVIK NATIONAL PARK, CANADA

The **Ivvavik National Park** in Yukon, Canada is a further example of a protected area in a tundra region. At 10 168 square kilometres, it is smaller than the ANWR and just across the US/Canadian border from it. It was set up to protect this area from possible oil developments and because it is a major caribou calving ground, with 169 000 caribou migrating through it. It was established in 1984 as a result of an agreement with the indigenous **Inuvialuit** people. Ivvavik is managed jointly by the Canadian government National Park Service and the Inuvialuit. Their aims in managing the park are:

- to protect and preserve the wilderness
- to look after wildlife (including polar and grizzly bears, wolves, wolverine, Arctic foxes, voles, lemmings)
- to preserve local cultures
- to allow the Inuvialuit to continue their subsistence lifestyle.

A shared vision for the Ivvavik National Park is: *'the land will support the people who protect the land'*. Visitors are welcomed but must apply for a permit before entering the park. To get into the park it is necessary to charter a small aircraft, buy a landing permit and pay a daily fee of about £16. The only other way into the park is to walk hundreds of kilometres over rough terrain. Once in the park visitors must be entirely self-sufficient as there are no facilities, camping grounds, marked trails or roads.

ENVIRONMENTAL CAMPAIGNS

Many environmental organisations work to raise public awareness of issues which threaten nature. Greenpeace is a well-known charity that has run many successful campaigns to stop activities which threaten the environment.

One such threat is currently the pressure to open up the Arctic for oil drilling, an activity that has become more likely due to the melting of the Arctic ice cap and the continuing world demand for fossil fuels. By taking direct action, encouraging the public to sign petitions, boycott offending companies and write to their political representatives, Greenpeace hopes to put pressure on oil companies and governments to persuade them to stop their activities in this sensitive environment, fearing that oil spills could cause irreparable damage to Arctic and tundra ecosystems. The Exxon Valdez disaster in Alaska in 1989 and continuing oil leaks in northern Russia are evidence of the threat from the oil industry. There is no doubt that the efforts of environmental groups, such as Greenpeace, prevent much environmental damage and ensure that commercial organisations behave in a more environmentally responsible way.

THINGS TO DO AND THINK ABOUT

1. At an international level, what things can be done to protect the Arctic and tundra environments?
2. For either the **Arctic National Wildlife Reserve** in Alaska or **Ivvavik National Park** in Canada, list the ways in which the tundra environment is protected.
3. How can an individual in Scotland help to protect remote tundra environments?

DON'T FORGET

The efforts of the UN to limit and control climate change are described on page 82.

Ivvavik National Park, Canada.

VIDEO LINK

Learn more about Greenpeace's campaign to save the arctic by watching the clip at www.brightredbooks.net/N5Geography

ONLINE TEST

Test yourself on managing tundra environments at www.brightredbooks.net/N5Geography

GLOBAL ISSUES

TROPICAL RAINFORESTS

Tropical rainforests are mostly found on or close to the equator. The main tropical rainforest regions are found in South America, Africa and south-east Asia. Key characteristics of tropical rainforests are:

- hot, humid climate
- high levels of precipitation
- no seasons
- dense vegetation
- huge variety of plant and animal species
- sparsely populated.

CLIMATE AND LOCATION

Tropical rainforest climate, also referred to as **equatorial climate**, is uniformly hot and wet. There is little difference from day to day. Often the difference between the night-time and daytime temperatures (the **diurnal range**) is greater than the annual temperature range. Average daytime temperatures are around 25–30°C with no or very little seasonal variation. Convectional rain occurs every day due to the humid air and intense insolation.

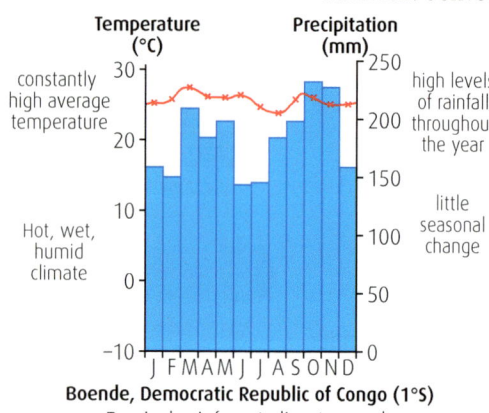
Boende, Democratic Republic of Congo (1°S)
Tropical rainforest climate graph.

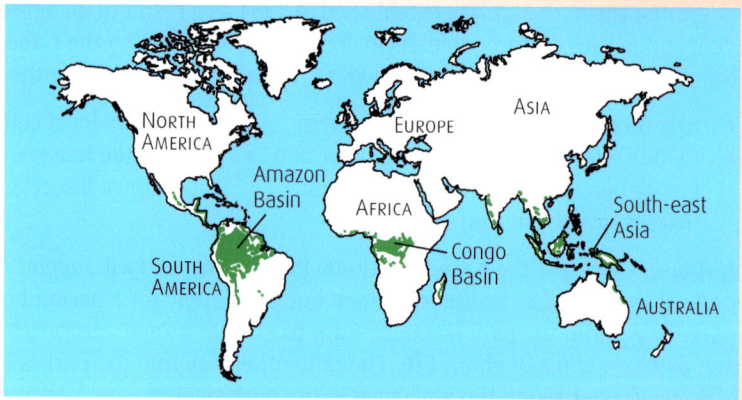
Diagram 3.11 Map of tropical rainforest areas.

TROPICAL RAINFOREST ECOSYSTEM

The tropical rainforest ecosystem is the most diverse on the planet. About 50% of known plant and animal species are found here. The all-year-round hot, wet conditions are ideal for plant growth, which can be very rapid. There are four distinct layers to the tropical rainforest. The very highest trees or **emergents** make up the top layer. They can grow over 50 metres high and appear above the next layer, known as the **canopy**. The canopy is a continuous platform of tree tops which blocks out the light from the forest below. The only gaps in the canopy are found where rivers flow through the forest. There are many different species of broad-leaved, evergreen trees but some of the most common include **tropical hardwoods** such as rosewood, teak, mahogany and ebony. **Epiphytes** are also common in the canopy. These are plants that support themselves by growing on others and deriving most of their nutrition from rain and the surrounding environment. Beneath the canopy is a layer of younger trees trying to break through the canopy as they compete for sunlight. This is known as the **under canopy**. As well as trees, there are **lianas**, vine-like plants which wind their way around tree trunks and greatly inhibit their growth. Much of the wildlife in the rainforest is found in the canopy and under canopy. At ground level is the **shrub layer**. Here there are ferns, seedlings and fungi, which help to break down the mass of decaying leaves, branches and tree trunks that have fallen from above. As little as 2% of the light available above the canopy reaches the forest floor so vegetation grows less densely than in the canopy. However, the forest floor is covered with rotting vegetation, so it is difficult for people to move around within the forest.

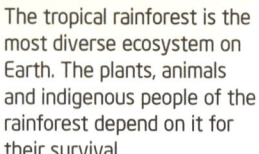

DON'T FORGET

The tropical rainforest is the most diverse ecosystem on Earth. The plants, animals and indigenous people of the rainforest depend on it for their survival.

contd

Global Issues – Tropical rainforests

Tropical rainforest vegetation is adapted to the climate in many ways. Plants grow all year round and so appear evergreen, although they continuously shed leaves while others grow to replace them. Leaves are broad to maximise the amount of sunlight they can obtain for use in photosynthesis and many have **drip tips** to help shed water during heavy downpours. Rainforest trees develop **buttress roots** to support their huge height and weight, while an extensive network of roots spreads out beneath the surface to help gather nutrients from the forest floor. This is an important part of the nutrient cycle, in which rapid decomposition of plant matter on the forest floor by bacteria, fungi and insects releases nutrients that can be quickly gathered by the root systems of surrounding trees. Forest soils are usually infertile, so the quick recycling of minerals and nutrients by plants and trees is vital for their survival.

Animals are most prolific in the canopy, where the dense network of branches creates aerial route ways for them in their search for fruits, berries, nuts and prey. Creatures such as orang-utans, spider monkeys, sloths, toucans and butterflies can be found here. Some of the larger forest animals are likely to be found on or closer to the forest floor, including jaguars (South America), tigers (Asia) and gorillas (Africa). However, there are many other smaller inhabitants of the forest floor, including frogs, snakes and creatures such as the leafcutter ant, which are equally important in the rainforest ecosystem.

Rainforest creatures are adapted to their environment in many different ways. Camouflage plays an important role. A tiger's stripes and a jaguar's spots allow them to blend in with the vegetation, while other creatures have developed the ability to change colour altogether such as the chameleon. Monkeys have prehensile tails allowing them to grip branches for support when climbing through the canopy. Birds, such as parrots, have sharp, strong and curved bills to crack open nuts, while their feet have two forward and two backward facing toes with elongated claws, allowing them to climb and grip branches for support and to manipulate their food easily. Their wings are short, allowing them to fly easily between branches and foliage in the canopy.

INDIGENOUS PEOPLES

It is estimated that there may still be up to 100 uncontacted groups of indigenous people in the rainforests of both the Amazon Basin and Papua New Guinea. However, the habitat and environment of many rainforest people has been very badly affected by uncontrolled deforestation. The largest indigenous groups of rainforest people are found in South America. The **Yanomami** people live in the Amazon rainforest, in the area of the Brazilian–Venezuelan border. The **Kayapo** inhabit the rainforest around the Xingu river, a tributary of the Amazon. In Africa, the **Baka** people follow a **subsistence** way of life in the Congo Basin. The **Penan** and **Dayak** are indigenous peoples of Borneo in south-east Asia.

Rainforest people have traditionally followed a **subsistence** lifestyle. The forest supplies all their needs, including building materials, clothing, weapons, food and medicines. They are hunter-gatherers but may also fish in local rivers and grow food in forest gardens. This is **shifting cultivation**, where small clearings are made by felling trees and burning the roots to add fertility to the soil. Crops, such as yams, sweet potatoes, papaya and manioc, are grown for several years before the clearing is abandoned because the soil has become infertile. A new clearing is made and the forest may grow back in the old one.

Indigenous forest people depend on the forest for their way of life and respect the integrity of the forest ecosystem. The Penan people of Borneo believe it is wrong to take more than what is necessary. Unfortunately, the culture and way of life of many forest people worldwide is increasingly under threat from other less ecologically aware forest activities.

Buttress roots and leaf drip tips.

Golden poison dart frog:
- brightly coloured to warn possible predators of danger
- secretes poison from special glands, making them toxic to eat
- feeds on poisonous ants
- has sucker pads on its toes allowing it to climb into the trees
- lays its eggs on leaves and flowers off the ground away from predatory fish.

 ONLINE TEST

Test yourself on tropical rainforests at www.brightredbooks.net/N5Geography

 VIDEO LINK

Watch the clip 'The global importance of rainforest photosynthesis' to learn more at www.brightredbooks.net/N5Geography

 THINGS TO DO AND THINK ABOUT

1. Explain how vegetation is adapted to the tropical rainforest environment. Try to give five different examples.
2. Make a star diagram to show how indigenous peoples of the tropical rainforest make use of their environment while following a subsistence way of life.

GLOBAL ISSUES

TROPICAL RAINFORESTS: CAUSES AND EFFECTS OF LAND DEGRADATION

BACKGROUND

For generations, tropical rainforests have been home to thousands of groups of indigenous people. They have not had a significant impact on the rainforest as their way of life is sustainable and very much in harmony with the surrounding forest. Indigenous forest peoples take only what they need from the forest and, for many groups, the trees and animals are sacred.

Unfortunately, commercial organisations and other groups of people not native to the forests have caused enormous damage to rainforests and the wildlife and indigenous peoples living in them. In the search for hardwood trees, minerals, oil, new land for farming, settlements and ranching, the world's rainforests have disappeared at an unprecedented rate. It is estimated by UNESCO that 50% of the world's tropical rainforest has disappeared in the last 50 years.

Before

- Plant nutrients used by trees in photosynthesis
- Falling leaves carry plant nutrients to forest floor
- Leaves decompose quickly in hot wet conditions
- Plant nutrients taken up by tree roots

After

- Supply of falling leaves removed
- Heavy rains erode soil and leach minerals away
- Soil is left barren and infertile
- Mineral cycle is broken

Diagram 3.12 The mineral cycle and deforestation.

THE MINERAL CYCLE

Soils in the rainforests are not fertile. The trees derive most of their minerals and nutrients from rapidly decomposing vegetation on the forest floor. Fallen leaves, twigs and branches are broken down quickly in the hot wet conditions, releasing plant nutrients into the top ten centimetres of the soil. These are quickly taken up again by the dense network of tree roots located just below the ground. When trees are felled and removed, this mineral cycle is broken. Without the supply of plant nutrients from falling leaves, the soil quickly become infertile. The heavy rains wash away any remaining nutrients in a process called **leaching**. Soon the land becomes barren and unable to support even basic agriculture. In shifting cultivation, practised by forest peoples, only small patches are cleared and the forest is allowed to grow back to some extent after three or four years of cultivation, with little lasting effect on the rainforest ecosystem. However, where the forest is removed permanently, leaching, soil infertility and soil erosion by heavy rains become major problems.

LAND DEGRADATION CAUSES

Deforestation has continued at an alarming rate. Tropical rainforests covered about 14% of the Earth's surface in the mid-twentieth century but now cover only 7–8%. Many estimates suggest that at the current rate of destruction all rainforests will be lost well before the end of this century.

The main causes of deforestation are:
- logging for hardwoods
- cattle ranching and commercial agriculture, including cultivation of **biofuels** such as soya and palm oil
- clearance for subsistence farming due to rapidly growing populations
- mining for minerals such as iron ore, bauxite, tin, nickel, copper and gold
- oil exploration and drilling
- the creation of hydro-electric power schemes and associated reservoirs (e.g. the Belo Monte Dam in Brazil will be the world's third largest hydro-electric power dam)
- road building.

Building roads through the forest is one of the most destructive forms of deforestation. Although only a small area of forest is cleared to build the road, it is the easy access that the roads create which leads to further destruction. Often illegal logging operations will spring up and these are very difficult to control in remote forest regions. People desperate for land will set fire to the forest to clear it and then claim the land as their own for farming. Fires often burn for days or weeks as there are few fire fighting resources in isolated areas of the forest.

DON'T FORGET

Deforestation has happened in all the world's forests but this section only describes deforestation in tropical rainforests.

contd

Global Issues – Tropical rainforests: Causes and effects of land degradation

Gold mining is one of the most destructive economic activities in the rainforest. Often it is uncontrolled and thousand of hectares of forest are destroyed by prospectors looking to make their fortunes. Powerful hoses are used to loosen soil from cleared areas of forest, causing irreparable soil erosion. Toxic substances, including mercury and cyanide, are used to separate the gold from other materials. These substances are washed into river systems, poisoning the water and killing wildlife further downstream. Indigenous people forced to drink and eat fish caught in these waters become ill and die. In Venezuela in the 1980s, gold mining in the Amazon had a devastating impact on the rainforest, its wildlife and the Yanomami people, who were often killed as they tried to protect their land from gold miners.

Gold mine at Yanacocha, Peru.

LAND DEGRADATION: EFFECTS

As rainforests are cut back, burned and destroyed the whole ecosystem is affected. Countless animals, insects and plants are killed and sometimes indigenous people too. As isolated rainforest tribes come into contact with outside groups, they are exposed to, and can die from, illnesses and diseases such as influenza and measles to which they have no natural resistance.

Many medicines used today are derived from rainforest plants. **Tuborcurarine** comes from the **curare** plant in Brazil. It is a muscle relaxant that is used in surgery and also in the treatment of multiple sclerosis. **Vincristine**, derived from the **rosy periwinkle** of the Malagassy rainforest in Madagascar, is used in chemotherapy, particularly in the treatment of leukaemia. **Quinine**, which is used in antimalarial treatments, is also derived from rainforest plants. These are three examples of plants with proven medicinal powers. Around 25% of cancer-fighting drugs originate from plants found in tropical rainforests, and there may be many other as yet undiscovered plant-based remedies perhaps known only to a few remaining indigenous groups. These species of plants could be lost forever. As yet undiscovered species of insects and animals may also be lost, while other endangered species could become extinct as their habitats are destroyed. On the Indonesian island of Sumatra, where rainforest has been cleared often for vast palm oil plantations, it estimated that there are as few as five to seven tigers left. They are likely to become extinct here.

- Indigenous peoples' land and way of life destroyed
- Devastating soil erosion
- Wildlife habitat destroyed
- Increased carbon dioxide levels contribute to global warming
- Rivers choked with soils and silt washed off by heavy rains

- Loss of animal species and possible extinctions
- Loss of potentially life-saving medicines from forest plants
- Water cycle affected: floods and drought
- Minerals and nutrients leached from soil surface

Diagram 3.13 Deforestation effects.

The loss of the tropical rainforest also affects the water cycle. Water is drawn up from the ground by root systems and captured on the surfaces of leaves. It is transpired from and evaporated from the leaves. Deforestation results in the loss of this natural cycle of water. The rainforest, which acted as a giant sponge, soaking up water, is no longer able to do this. Water runs off the land rapidly, causing devastating soil erosion and flash floods. Droughts can also result from large-scale deforestation as less water enters the atmosphere due to reduced evapotranspiration.

Tropical deforestation is a significant contributor to the enhanced greenhouse effect, contributing to global warming. As trees are removed, they are no longer able to absorb CO_2 from the atmosphere and, where the rainforest is burned, additional CO_2 is added to the atmosphere. It is estimated that this contributes 1.5 billion tonnes of CO_2 to the atmosphere each year, around 12% of the total annual global CO_2 emissions.

The continuing loss of tropical rainforests is one of the most important environmental issues of our time. Deforestation has wide-ranging local and global impacts and it is imperative that solutions are found to halt this destruction. About 1.5 hectares of tropical rainforest are destroyed every second. International, as well as local, efforts are being made to prevent further needless destruction, but so far with only limited effects.

DON'T FORGET

Deforestation is a major cause of global warming.

ONLINE TEST

Test yourself on tropical rainforests at www.brightredbooks.net/N5Geography

VIDEO LINK

For more about the effects of deforestation on tropical rainforests; watch the video clip at www.brightredbooks.net/N5Geography

THINGS TO DO AND THINK ABOUT

1. Explain how deforestation of tropical rainforests leads to:
   (a) soil infertility and erosion
   (b) global warming.

2. Make a star diagram to show how deforestation affects indigenous peoples such as the Yanomami of Venezuela.

GLOBAL ISSUES

TROPICAL RAINFORESTS: SOLUTIONS TO DEFORESTATION

BACKGROUND

Deforestation is devastating to the local rainforest environment. With clear felling, where all the trees are removed, the ecosystem is totally destroyed. Animals and plants are killed, many endangered species are put at even greater risk of extinction as their habitat shrinks and indigenous people lose their land, culture and entire way of life. There are global implications too, as the balance of CO_2 and oxygen in the atmosphere changes and the amount of water vapour transpired from vegetation is reduced, both of which can lead to climate change.

Managing deforestation is very difficult as the areas most at risk are often remote and inaccessible. Fires burn uncontrollably and their presence may sometimes only become apparent because of satellite photographs. Trees are often felled illegally but many equatorial countries do not have the resources to track down and prosecute those responsible.

PREVENTING DEFORESTATION

Every year an area of 143 000 square kilometres of tropical rainforest is destroyed. That is almost twice the total land area of Scotland!

Governments in countries with tropical rainforests have introduced laws making much of the deforestation illegal, but often these laws are ineffective and difficult to enforce. In 2004 the government in Paraguay successfully introduced a zero deforestation policy and in the following years deforestation was reduced by 85%. Many environmental organisations (e.g. Greenpeace, the Rainforest Alliance and the WWF) work with governments and put pressure on logging companies. One aim is to have a zero deforestation policy introduced by all countries with tropical rainforests by 2020.

Current laws regulate most aspects of logging in tropical rainforests. However, laws are often broken when logging companies take more timber than is permitted by harvesting trees from officially protected areas or by taking protected species of tree. Sometimes, illegal logging operations account for more timber than legally-sanctioned felling. This deprives governments of taxes and revenues which come from legal forestry operations, lowers the price of timber internationally and disadvantages law-abiding companies, who find it difficult to compete with more cheaply produced illegally sourced timber.

Another strategy to stop illegal logging is the setting up of protected forest areas. This prevents the destruction of forest ecosystems, preserving endangered plant and animal species for future generations. The **Virunga National Park** is in the eastern mountains of the Democratic Republic of Congo. It borders national parks in Uganda and Rwanda, giving a large area of protected rainforest where critically endangered mountain gorillas have been able to survive. Although poaching is an ever-present threat, the Virunga National Park is an example of a conservation measure that has succeeded in protecting an area of rainforest and its diverse ecosystem. However, current proposals to drill for oil there pose a further environmental threat. One oil company has already abandoned its drilling plans because of concerns that environmental damage would harm its reputation.

In the Amazon Basin, a programme of conservation was introduced in 2003 to protect the tropical rainforest and its ecosystem. Known as the **Amazon Region Protected Areas (ARPA)** programme, it is part-funded by environmental organisations, the Brazilian government and the UN. **Tumucumaque Mountains National Park** in Brazil is part of the ARPA programme. At almost 39 000 square kilometres, the park is the same size as Switzerland. It borders a similar area of protected land in French Guiana, the **Guiana Amazonian National Park**, together forming one of the largest areas of protected rainforest in the world. As well as preventing logging operations and deforestation, the Tumucumaque Mountains National Park helps to protect a variety of endangered species such as jaguars, primates, macaws, harpy eagles and turtles.

DON'T FORGET

Ecotourism is another way of slowing down deforestation. Tourists want to experience the sights, sounds and wildlife of the rainforest ecosystem. By providing undisturbed reserves of tropical rainforest, to which visitors are prepared to pay for entry, money can be raised for projects that help to further protect the forest.

Global Issues – Tropical rainforests: Solutions to deforestation

SUSTAINABLE FOREST MANAGEMENT

A number of forest products can be produced in a **sustainable** way, that is by not damaging the rainforest or its biodiversity, leaving it undiminished for future generations. Brazil nuts, for example, can be collected from trees in the Amazon without causing any damage to the surrounding environment. Rubber can also be collected by tapping trees growing in the tropical rainforest. This is done while trees continue to grow and allows the rainforest to flourish round about.

Producing timber from sustainably managed forests is much more difficult. Timber that is produced 'sustainably' may simply mean that there will continue to be a supply of that type of timber in the future. For hardwood timber production to be truly sustainable, no rainforest should be destroyed during harvesting. This can be achieved by felling single trees and removing them from the forests using a minimum amount of machinery (often ox-drawn carts) so that only the narrowest of strips are cleared for access, allowing natural forest regeneration to take place afterwards.

Education is key to protecting rainforest areas from clearance by local people. If they are taught about the advantages of sustainable forest management, there is more chance that tropical rainforests will be preserved for future generations.

Isolated communities in areas of tropical rainforests understandably want to use the rainforest and its resources for their own benefit. Often sustainable forest management involves a compromise between the needs of the local community and the desirability of preserving the rainforest intact. Environmental organisations and governments work hard to allow communities to make the most of local resources while still protecting the vulnerable rainforest.

Using oxen instead of machines to remove logs reduces damage to the forest.

ONLINE TEST

Test yourself on tropical rainforests at www.brightredbooks.net/N5Geography

VIDEO LINK

Watch the video 'Tropical Rainforests: The Affordable Climate Change Solution' to learn more at www.brightredbooks.net/N5Geography

AFORESTATION

Fully restoring rainforest ecosystems that have been destroyed by deforestation cannot be achieved easily or quickly. The easiest areas to restore are those that have been used for slash and burn agriculture, where forests have been burned and used for small-scale agriculture often on a subsistence basis. If this land is close to the rainforest and is left to regenerate, trees will quickly establish themselves, providing a thin canopy to protect the soil. Over the longer term, the forest might grow back. However, farmers have to be persuaded to leave their land and are only likely to do this if there is an alternative. In recent years, a scheme called **Reducing Emissions from Degradation and Deforestation (REDD+)** has involved wealthy countries funding schemes in developing countries that involve restoring damaged forest ecosystems. This might involve finding new livelihoods for famers who have been involved in slash and burn agriculture, perhaps using degraded grassland instead of rainforest or providing health care to communities who depend on illegal logging to pay for medicines. Countries such as Australia or Germany might fund projects to help restore rainforests or reduce deforestation in countries such as Papua New Guinea or Ecuador.

- Environmental organisations highlight destructive practices and put pressure on companies and governments to stop them
- Encourage customers to buy wooden products which are from sustainable sources
- Island biogeography
- Establishing reserves and national parks (e.g. Virunga National Park, Congo)
- Revenue from ecotourism can be used for rainforest friendly projects
- Introduce and enforce strict conservation laws to prevent widespread logging
- Encouraging companies only to buy timber from sustainable sources (e.g. Global Forest and Trade Network (GTFN))
- Strip harvesting reduces damage and allows forest to regenerate

Diagram 3.14 Solutions to deforestation.

ONLINE

Check out the Greenpeace guide on solutions to deforestation at www.brightredbooks.net/N5Geography

THINGS TO DO AND THINK ABOUT

1. Give reasons to explain why preventing deforestation is difficult.
2. List different examples of how tropical rainforests can be sustainably managed.

GLOBAL ISSUES

ENVIRONMENTAL HAZARDS: VOLCANOES, EARTHQUAKES AND THEIR CAUSES

The Earth's crust is made up of a series of plates. Oceanic crust is usually thin and dense, while continental crust is thicker but less dense. Sometimes crustal plates can be made up of both oceanic and continental crust. Earthquakes and volcanoes result mostly from movements of the Earth's crustal plates.

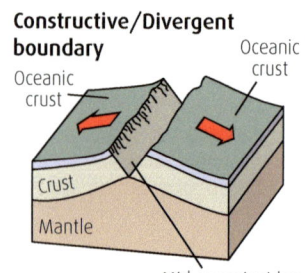

PLATE BOUNDARIES

Most earthquakes and volcanic activity occur close to, or on, the plate boundaries. There are four different types of plate boundary:

- constructive plate boundaries
- destructive plate boundaries
- collision plate boundaries
- conservative plate boundaries.

Constructive plate boundaries are where two oceanic plates are moving apart. This causes a weakness in the Earth's crust and makes volcanic eruptions more likely. Often these plates can be in the form of a line of volcanoes, as is found along the Mid-Atlantic ridge. Ascension Island, the Azores and Iceland are all volcanic islands which owe their existence to volcanic activity along the Mid-Atlantic ridge. As the plates move away from each other, volcanic eruptions spew out lava, which solidifies to form new crust.

At a **destructive plate boundary**, where the oceanic and continental crusts collide, one plate is forced beneath the other in what is known as a **subduction zone**. The friction caused by the collision of the two plates often stops them from moving further until, as tension builds up, they suddenly jolt apart, causing a major earthquake. The subducted oceanic plate may melt due to intense friction, heat and pressure, creating **magma**. This magma, which is less dense than the surrounding rocks, forces its way to the surface, creating volcanoes. The Andes is a chain of fold mountains caused by the continued collision of the Nazca and South American plates. A series of volcanoes, such as Nevado del Ruiz in Colombia and Cotopaxi in Ecuador, have been formed, and the area is subject to violent earthquakes such as in Chile in 2010 when a subsea earthquake reaching 8.8 on the Richter scale caused a huge earthquake and tsunami.

At a **collision plate boundary**, two continental plates collide, causing the crust to buckle and form fold mountains. This mountain building is known as orogenesis.

Earthquakes can occur here although, because there is not a subduction zone, volcanic activity is less likely. The Himalayas, the world's greatest fold mountain range, has been created where the Indo-Australian plate collides with the Eurasian plate.

A **conservative plate boundary** is where two plates slide past each other and crust is neither created nor destroyed. Often friction causes tension to build up and large earthquakes can occur. The San Andreas fault is a conservative plate boundary where the Pacific and North American plates are moving parallel to one another.

Diagram 3.15 Plate boundaries.

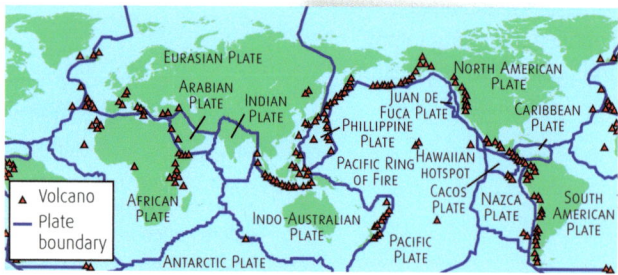

Diagram 3.16 Worldwide location of volcanoes.

VOLCANOES

Volcanoes occur along plate boundaries, although there are some that occur on so-called 'hot spots' away from plate margins. Hawaiian volcanoes, such as Muana Loa, are an example of this. Around the margins of the Pacific and Nazca plates where they meet continental plates lies the Pacific Ring of Fire, a particularly active ring of hundreds of volcanoes which encircles the Pacific Ocean.

contd

Global Issues – Environmental hazards: Volcanoes, earthquakes and their causes

An **extinct** volcano is one that will never erupt again, while a **dormant** volcano is one that has not erupted for a long time but could still do so. **Active** volcanoes erupt regularly, such as Mount Etna in Sicily, which can erupt several times a year.

A typical volcanic cone is made up of a central pipe or **vent** leading from the **magma chamber** to the **crater** at the top, from which eruptions may happen. Sometimes this may be blocked by solidified **lava** known as a **plug**. As pressure builds from the magma below, a violent eruption can blow the plug off, leading to an explosion of **volcanic ash**, dust and **poisonous gas** that can rise high into the atmosphere. **Volcanic bombs** are lumps of semi-molten rock that can be catapulted several kilometres from the volcano, while molten rock emitted from the crater runs down the volcano's side as a **lava flow**. The most devastating volcanic eruptions may result in a **pyroclastic flow**, a lethal mix of hot rocks, volcanic ash and gases that travels very rapidly under force of gravity killing anything in its path. Sometimes, there may be a secondary vent and crater on the side of the volcano known as a **secondary** or **parasitic cone**. A vent from which sulphurous gases are emitted is known as a **fumarole**.

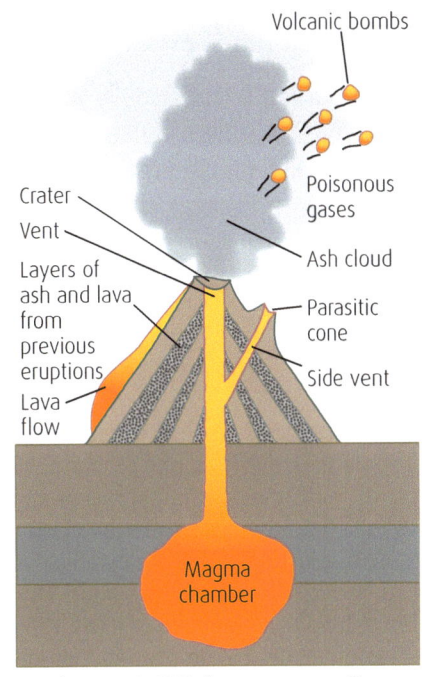

Diagram 3.17 Volcano cross-section.

Ngauruhoe: An active cone-shaped volcano in New Zealand.

 DON'T FORGET

Plate boundaries are fault lines that separate the Earth's tectonic plates. One of the most notorious fault lines is the San Andreas fault in California.

VIDEO LINK

Check out the volcanoes clip to learn more at www.brightredbooks.net/N5Geography

EARTHQUAKES

Most earthquakes occur on, or close to, plate boundaries. Although there are hundreds of earthquakes every day, most are very small and are not noticed by people in the area. The severity of an earthquake is measured on the **Richter scale**, which is logarithmic, with a force 6 earthquake being 10 times stronger than a force 5 earthquake. The highest recorded earthquakes have not exceeded force 9. The 2011 Tohoku earthquake off the Japanese coast measured 9.0 and generated a devastating tsunami.

Earthquakes are caused by the movement of rocks along the edges of the Earth's crustal plates. The point beneath the surface where the movement takes place is known as the **focus**, and earthquake shockwaves radiate outwards from here while the point directly above this on the surface is known as the **epicentre**. Damage is worst close to the epicentre. There are three types of earthquake waves:

- P waves (primary or pressure waves), which move rocks up and down and travel the fastest
- S waves (secondary), which move the rocks from side to side and travel at 60% of the speed of P waves
- L waves (long waves), which travel along the Earth's surface in a similar way to waves on the ocean surface. These are the slowest waves but because of their rippling effect they cause the greatest damage.

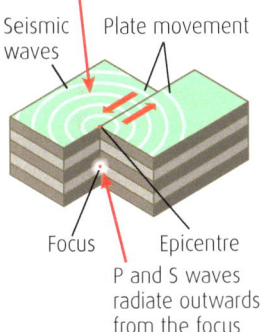

Diagram 3.18 Earthquake features.

 VIDEO LINK

Watch the clip about earthquakes to learn more at www.brightredbooks.net/N5Geography

 THINGS TO DO AND THINK ABOUT

1. Explain why volcanic activity is particularly likely at destructive and constructive plate margins.
2. At the boundaries of which plates are these famous volcanoes found?

    (a) Mount Fuji
    (b) Mount Etna
    (c) Cotopaxi
    (d) Mount St Helens

 ONLINE TEST

Take the test on volcanoes and earthquakes at www.brightredbooks.net/N5Geography

97

GLOBAL ISSUES

ENVIRONMENTAL HAZARDS: VOLCANOES – EFFECTS AND MANAGEMENT

The effects of volcanic eruptions are not usually experienced as widely as the effects of large earthquakes. Occasionally, an ash plume may cause widespread effects, as in Europe in 2010 when the Icelandic volcano Eyjafjallajökull caused massive disruption to air transport as volcanic ash spread over northern and western Europe, making it unsafe for aircraft to fly. There are several threats to the surrounding area during volcanic eruptions. These are lava flows, ash clouds, volcanic bombs, poisonous gases, pyroclastic flows and lahars.

MOUNT MERAPI: AN ACTIVE VOLCANO

Eruption of Mount Merapi, Indonesia. A hot ash cloud and pyroclastic flow rapidly descended its slopes, killing people up to 13 kilometres away.

Mount Merapi, Indonesia's most active volcano, is situated in the centre of its most populous island, Java, and reaches 2930 metres above sea level. It has erupted frequently for over 400 years. Its most recent major eruption started on 23 October 2010 when more than 500 earthquakes were recorded on the mountain that weekend. An evacuation zone of ten kilometres was established by the Indonesian government as villagers were ordered to leave for safer ground. However, following the first eruptions, which spread hot ash up to 13 kilometres away, the evacuation zone was extended to 15 and then 20 kilometres, displacing over 400 000 people. Numerous volcanic eruptions took place over the following three weeks, with burning clouds of hot ash, gas and pyroclastic flows descending suddenly from Merapi and killing 353 people. The message to evacuate was ignored by some villagers who refused to abandon their houses, livestock and farms, worried about losing all their possessions. Others had lived through previous eruptions and saw no need to evacuate. Unfortunately, Merapi claimed the lives of many people due to severe burns but mostly due to asphyxiation as the fine ash and poisonous gas filled their lungs. An ash cloud and a plume of sulphur dioxide extended for thousands of kilometres over the Indian Ocean following the eruptions.

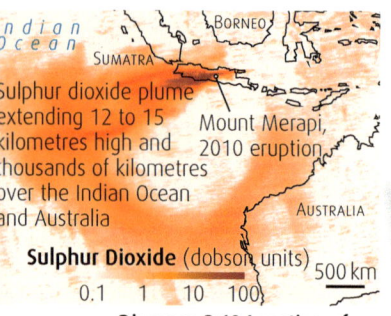

Diagram 3.19 Location of the sulphur dioxide plume following the 2010 Merapi eruption.

AID EFFORTS FOLLOWING THE ERUPTION OF MOUNT MERAPI

Following the first eruptions, rescue teams of soldiers and volunteers helped locate the dead and injured. Emergency medical supplies, as well as drinking water and face masks, were shipped into the area as everything was covered in ash.

Emergency relief centres were established for evacuees and refugees where aid organisations, such as the Indonesian Red Cross, distributed food, water, sleeping mats and medication. Field hospitals were established where teams of emergency medical workers treated victims suffering from burns and breathing difficulties following the eruptions. Foreign aid organisations and countries such as the United Arab Emirates funded these rescue efforts. Government aid was needed for local communities to rebuild schools and health care facilities and to repair electricity and water supply networks. The **Asian Development Bank** gave an initial grant of $3 million dollars to help villagers repair damage and to allow farmers to restock (crops were destroyed and farm animals killed by hot ash clouds). This is an example of **multilateral aid**, where grants are given by an organisation funded by many different countries.

Relief centre for evacuees: volunteers of the Indonesian Red Cross distribute food, water and plastic mats to tens of thousands of people fleeing Mount Merapi.

MOUNT MERAPI: REDUCING THE RISK

Java is the most densely populated island in Indonesia, famous for its rich and fertile volcanic soils. A combination of good farmland and shortage of land results in thousands of people settling in potentially hazardous areas close to active volcanoes. Indonesia is on the Pacific Ring of Fire and is one of the most active volcanic and earthquake zones in the world.

contd

Global Issues – Environmental hazards: Volcanoes – effects and management

Government scientists closely monitor Mount Merapi and other active volcanoes for imminent signs of eruption. These include bulges caused by an upwelling of magma close to the surface around the base and slopes of volcanoes, which can be detected by satellite imaging. Seismometers detect minor earthquake activity, which can increase before eruptions and give valuable time for alerts to be issued to surrounding communities. The city of Yogyakarta, with a population of 500 000 people, lies just 30 kilometres south of Mount Merapi and would be at risk if ash from a major eruption was blown southwards. Evacuation plans exist for the areas surrounding the volcano, with different categories of evacuation zones according to the severity of any potential eruption. A further danger at Merapi is the possibility of **lahars**, a combination of mud, ash, lava and water that can flow very rapidly down river valleys, burying surrounding houses and farmland. In some villages close to the volcano, protective barriers have been constructed around important buildings, such as primary schools, to divert any lahars away from them.

ONLINE

For more on this, read the BBC news report on the Mount Merapi eruption in 2010 at www.brightredbooks.net/N5Geography

DON'T FORGET

Lahar is a Javanese word meaning a volcanic mudflow. Lahars can be devestating. In 1985 lahars from the volcano Nevado del Ruiz in Columbia, buried the town of Armero, killing an estimated 23 000 people.

BENEFITS OF VOLCANOES

It is easy to overlook the positive side of volcanic activity, which has many benefits for local communities and their economies. Volcanic soils are rich and very fertile, making volcanic slopes and the surrounding areas attractive areas to farm and settle in (e.g. the Indonesian island of Java, home to 136 million people, the most populous island in the world, famous for its rich rice lands). Volcanic mud is used in cosmetics because of its cleansing and purifying effect on skin. Volcanic activity is also a tourist attraction, with local economies benefiting from thousands of visitors each year who travel to see the raw beauty of nature (e.g. Mount Etna, Sicily and also Iceland). **Geothermal power** is a major benefit, with volcanic regions such as Iceland and North Island in New Zealand generating electricity this way. Precious stones, such as diamonds, and metals such as gold and silver are found in areas that have been subject to volcanic activity (e.g. South Africa). New land can be created by lava flows when they meet the sea. The town of Heimaey on the Westmann Islands in southern Iceland famously benefited from a new, more sheltered harbour thanks to the eruption of the volcano Eldfjell in January 1973.

DON'T FORGET

Predicting volcanic eruptions is not an exact science. Sometimes eruptions can be unexpected and much larger than predicted. Mount St Helens was closely monitored but still killed 61 people when it erupted in May 1980 due to the severity and direction of the eruption.

| Fertile volcanic soils |
| Poisonous gas |
| Geothermal power |
| Pyroclastic flows |
| Volcanic mud used to cleanse and purify skin |
| Lahars |
| Precious stones and metals |
| Volcanic ash clouds |
| New land |
| Volcanic bombs |
| Visiting tourists create jobs and benefit local economies |

Diagram 3.20 Volcanic hazards and benefits.

| Satellite imaging detects bulges caused by magma close to surface |
| Seismometers detect increased earthquake activity prior to eruptions |
| Emergency services on alert; medical supplies as well as face masks and bottled water stockpiled |
| Warnings issued; exclusion zones established; people evacuated |
| Temporary and permanent barriers constructed in areas at risk from lava |
| Windows of buildings in areas at risk from lava bombs boarded up |

Diagram 3.21 Managing the impact of volcanic hazards.
Source: http://www.volcanoinfo.co.uk.

THINGS TO DO AND THINK ABOUT

1. What are the main hazards of volcanic eruptions and how can they affect surrounding communities?
2. For the eruptions of Mount Merapi in 2010 list:
   (a) the main effects
   (b) the main rescue efforts.
3. Write down the main benefits of volcanoes and for each benefit give an example of a place where this is experienced.

ONLINE TEST

Take the test on volcanoes at www.brightredbooks.net/N5Geography

GLOBAL ISSUES

ENVIRONMENTAL HAZARDS: EARTHQUAKES – EFFECTS AND MANAGEMENT

Earthquakes are comparatively common, although most are barely felt or happen in areas where there are few people. Occasionally, a larger earthquake happens in a more densely populated area and that is when widespread damage occurs. Earthquake shock waves are transmitted through the Earth and along the Earth's plates. The shaking and rippling effect of these waves can destroy buildings and in coastal areas create devastating tsunamis, as with the 2011 Tohoku earthquake in Japan. The extent of damage and the number of casualties can be far higher than in a major volcanic eruption.

CHILEAN EARTHQUAKE 2010

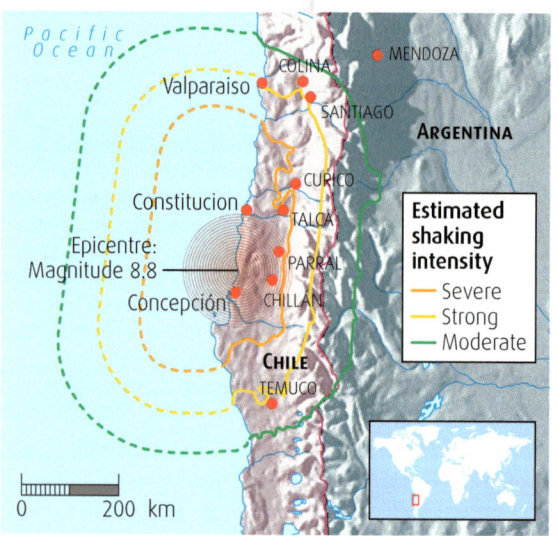

Diagram 3.22 Chilean earthquake, 27 February 2010.

At 3.30am on 27 February 2010 a magnitude 8.8 earthquake took place on the Chilean coast. The earthquake happened where the Nazca plate is being pushed beneath the South American plate in a subduction zone along the west coast of South America, part of the Pacific Ring of Fire. This was a huge earthquake that was felt as far away as Buenos Aires on the east coast of the continent. Many buildings in the cities of Concepcion and Chillan, 60 kilometres from the epicentre, were destroyed or badly damaged. In total, some 525 people were killed and 12 000 injured as a result of the earthquake. Serious damage was caused 335 kilometres away in the capital Santiago, where 13 people were killed as several buildings collapsed, including a newly constructed apartment block that was supposedly earthquake proof. The international airport was closed for three days due to the effects of the earthquake. Altogether, over 1.5 million people were displaced as their houses were destroyed, along with over 4000 schools and 79 hospitals. Telecommunications links, and water and electricity supplies were badly disrupted and many bridges destroyed. A major tsunami warning was issued in the Pacific following the earthquake, with two-metre high tsunamis affecting French Polynesia and many other Pacific nations. A number of people were killed, harbours were damaged and numerous boats were destroyed by the resulting tsunami, which reached two metres in height at Valparaiso, west of Santiago. In remote coastal villages, reports described upturned boats and cars side by side on land. The fear of aftershocks resulted in people sleeping outside for weeks afterwards rather than risk staying in their homes. Eight aftershocks were recorded within 24 hours of the main earthquake, with the largest of these being 6.9 in magnitude. Within a month there had been over 100 aftershocks.

DON'T FORGET

There is often no warning of an earthquake. There is anecdotal evidence that animals may behave unusually immediately before an earthquake but this has not been scientifically proven. Earthquake prediction is therefore difficult. Most efforts focus on limiting the impact of earthquakes.

CHILEAN EARTHQUAKE 2010: RESCUE EFFORTS

Immediately following the earthquake, rescue workers were sent to the worst affected areas to help victims and those still trapped in the rubble. Infra-red heat-seeking cameras were used as well as sniffer dogs to help locate people inside damaged buildings. Looting broke out in many towns following the earthquake as law and order threatened to collapse. However, police and 14 000 soldiers eventually restored order, although this took several days and involved curfews and the use of water canon to break up mobs of looters, among other measures.

Chile requested aid from other countries. Many foreign countries, such as the USA, Canada, the European Union (EU), China and Argentina, sent multi-million pound aid

contd

Global Issues – Environmental hazards: Earthquakes – effects and management

shipments early in March. These included field hospitals, rescue teams, tents, blankets, food, temporary shelters, portable electricity generators, bottled water and water-treatment plants. International aid agencies, such as the Red Cross, sent medical supplies and medical workers in the days immediately following the earthquake. All of these are examples of **short-term aid**, aimed at restoring basic services and living conditions as quickly as possible. **Long-term aid**, often received in the form of loans, has helped to restore infrastructure and to rebuild schools, houses and hospitals, although this process is likely to continue for many years.

Earthquake and tsunami damage in the Chilean town of Concepcion.

LIMITING THE IMPACT OF EARTHQUAKES

In total, the Chilean government estimates it will cost at least £20 billion and take five years to repair the damage caused by this earthquake. Many scientists now believe that it might not be possible to accurately predict precisely when major earthquakes will occur and so the best option to reduce the number of casualties is to construct buildings that are more able to resist earthquake damage.

There are a number of different ways in which buildings can be made more resistant to damage from earthquakes, including:

- constructing buildings that are effectively supported on rollers and will absorb energy from L waves (e.g. the Yokohama Landmark Tower, Japan)
- incorporating a counterweight at the top of buildings to reduce the damaging swaying motion caused by earthquake shock waves (e.g. Taipei 101, Taiwan)
- incorporating giant shock absorbers into building designs, as with Torre Mayor, Mexico City's 57 storey skyscraper, which lies close to the epicentre of a previous large earthquake and is designed to resist earthquakes of magnitude 9.0 (see Diagram 3.23).

Diagram 3.23 Torre Mayor, an earthquake-proof building in Mexico City.

Other measures to reduce the impact of earthquakes include regular earthquake drills held in countries such as Japan and the Philippines. In Japan, situated where three different tectonic plates meet, National Disaster Prevention Day is held on 1 September each year, the anniversary of the Great Kanto Earthquake in 1923, which claimed the lives of 140 000 people in Tokyo. Schools practise their earthquake drills in which pupils shelter beneath desks and tables to protect their heads during an earthquake and evacuate the building immediately afterwards. There are also earthquake simulators where people can learn how to react in an earthquake. As part of emergency planning, Japanese residents are encouraged to maintain an emergency box containing items, such as bottled water, purification tablets, tinned food and torches, which could be vital following disruption caused by a major earthquake.

Tsunami warning systems have been developed so that people in coastal areas can be alerted following a major earthquake that might have happened thousands of kilometres away. In the 2004 Asian earthquake 230 000 people were killed, mostly due the effects of the resulting tsunami. As a result, the Indian Ocean Tsunami Warning System was established, similar to one which already existed in the Pacific Ocean. In theory, coastal residents will have time to escape to higher ground when a tsunami alert is issued.

VIDEO LINK

Learn about the 1995 Kobe earthquake in Japan by watching the clip at www.brightredbooks.net/N5Geography

THINGS TO DO AND THINK ABOUT

1. What were the main effects of the Chilean earthquake on 27 February 2010?
2. What sort of help is required following a major earthquake such as this?
3. List all the ways in which the impact of earthquakes can be reduced.

ONLINE TEST

Take the test on earthquakes at www.brightredbooks.net/N5Geography

GLOBAL ISSUES

ENVIRONMENTAL HAZARDS: TROPICAL STORMS

A **tropical storm** is a severe low pressure system with sustained wind speeds of over 60 kilometres per hour. They form in tropical or sub-tropical oceans where the surface temperature of the water is at least 26°C. Tropical storms become **hurricanes** when sustained wind speeds reach 120 kilometres per hour or more. Each year, an average of 86 tropical storms and 49 hurricanes inflict damage and destruction on those areas which lie in their path.

FORMATION

Tropical storms form when the sea surface temperature reaches at least 26°C and evaporation causes strong uplift of moist air. Air rises because it is hot and lighter than the surrounding air. This creates low pressure at the surface. When warm moist air rises rapidly it cools and condenses, forming cumulonimbus clouds. As the air continues to rise, it starts to spread out and spiral round because of the Earth's rotation. As more and more air rises, thicker clouds form, giving heavy rain and more air is sucked in at sea level to replace the rising air. The air being drawn across the surface of the ocean can create very strong winds due to the large pressure difference between the centre of the storm and the surrounding area. The fastest speeds are often recorded around the edge of the centre or **eye** of the storm. A fully developed tropical storm can have a diameter of 1000 kilometres or more, causing its effects to be spread over a very large area. As tropical storms can last for a week or more, they affect huge areas and many different countries as they develop and move.

VIDEO LINK

Watch the video clip about the formation of a tropical storm at www.brightredbooks.net/N5Geography

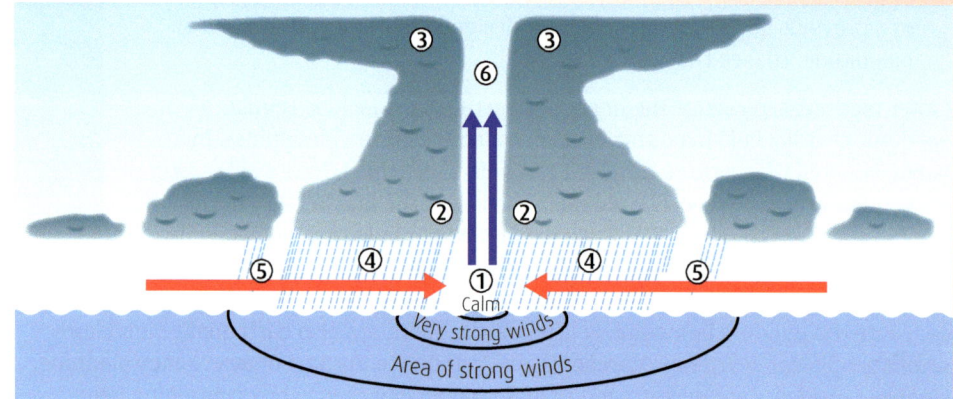

Diagram 3.24 Tropical storm cross-section.
1. Warm, moist air rises;
2. Water vapour condenses, forming cumulonimbus clouds;
3. Uplift continues and air spreads out and starts to spiral;
4. Heavy rain;
5. Air drawn in towards the centre of the storm creates very strong winds;
6. Eye of the storm.

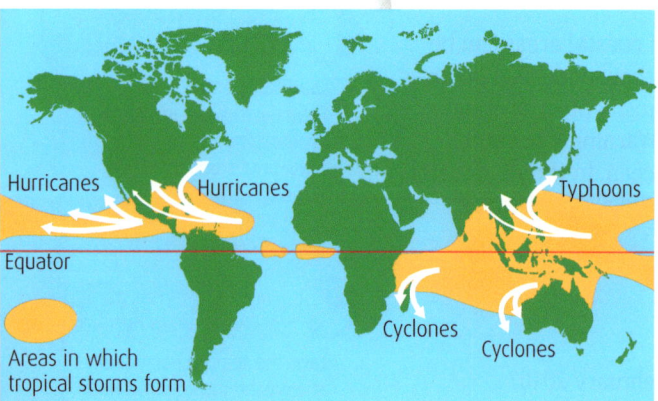

Diagram 3.25 Worldwide locations of tropical storms.

FEATURES OF HURRICANES

In the centre of the storm, at the eye, it is calm and skies may briefly be clear as the clouds and wind spiral around it in a huge vortex. The very strongest winds occur around the edge of the eye, where gusts in excess of 380 kilometres per hour (236 mph) have been recorded. Winds of this strength are a threat to life and nearly all buildings. Torrential rain is associated with tropical storms. Affected areas can experience a month's rainfall in a few hours, leading to severe flooding and landslides. Coastal areas often experience a **storm surge**, where a combination of extreme low pressure and fierce onshore winds can temporarily raise sea levels several metres above the normal high tide mark. This can also lead to devastating floods, as in New York City during Super Storm (Hurricane) Sandy in October 2012 when parts of the city's underground system were flooded and large areas of the city lost power.

Global Issues – Environmental hazards: Tropical storms

LOCATION OF TROPICAL STORMS

Severe tropical storms are known as hurricanes in the North Atlantic and Caribbean. They are called **cyclones** in the Indian Ocean and **typhoons** in the Pacific Ocean and South China Sea. It is rare for tropical storms to form in the southern Atlantic or Pacific. These storms normally move or track in a westerly direction but can sometimes become stationary or stalled for many hours, inflicting sustained damage on areas unlucky enough to be affected. As storms **dissipate** or die out, they may move north-easterly in the northern hemisphere and south-easterly in the southern hemisphere. As they move away from tropical and sub-tropical regions, there is no longer enough warm surface water to drive them so they become less intense. However, storms which were originally Atlantic hurricanes do sometimes track all the way across the Atlantic, arriving on the shores of northern Europe to cause considerable disruption. The strength of hurricanes and tropical storms is measured on the **Saffir–Simpson** scale as shown in the table.

A depression becomes a tropical storm if wind speeds are sustained between 63 and 118 km/h. A tropical storm becomes a hurricane once sustained wind speeds reach 119 km/h (74 mph) or more. There are five categories of hurricane, with 1 being the least damaging and 5 being the most severe. Typhoons and cyclones are also given categories using the same scale. Typhoon Usagi (2013) in south-east Asia and Hurricane Sandy in North America (2012) were both category 3 hurricanes. Thankfully, category 5 hurricanes do not happen every year. Hurricane Katrina (2005), which caused so much devastation in New Orleans, was an example of this most devastating level of hurricane.

Hurricane categories

Category	Sustained wind strength
5	over 252 km/h (157 mph)
4	209–251 km/h (130–156 mph)
3	178–208 km/h (111–129 mph)
2	154–177 km/h (96–110 mph)
1	119–153 km/h (74–95 mph)
Tropical storm	63–118 km/h (39–73 mph)

Satellite photograph of a hurricane.

TROPICAL STORM NAMES

Tropical storms are most likely to occur when the water temperature on the surface of the ocean is at its warmest. This is usually at the end of the summer but the official hurricane season in the North Atlantic starts on 1 June and finishes on 30 November. There are many tropical storms during each season and it is not unusual for two or three storms or hurricanes to be active in one ocean area at the same time. To make it easier for people such as forecasters, emergency response workers and ships' captains, the World Meteorological Organisation makes up a list of names, assigned in alphabetical order, for tropical storms as they develop during each hurricane season. If a tropical storm becomes a hurricane, it keeps the name already assigned to it as a tropical storm. Names may be repeated after six years but the names of the most damaging hurricanes (e.g. Katrina, Sandy) are not used again. The table shows the list of names for tropical storms for 2015.

Names for Atlantic tropical storms for 2015

Ana	Bill	Claudette
Danny	Erika	Fred
Grace	Henri	Ida
Joaquin	Kate	Larry
Mindy	Nicholas	Odette
Peter	Rose	Sam
Teresa	Victor	Wanda

 DON'T FORGET

The eye of a hurricane can be up to 50 kilometres (over 30 miles) across. As the eye passes over, the fierce winds and torrential rain can suddenly stop to be replaced briefly by blue skies and sunshine before the strong wind and rain begin again on the other side of the hurricane and from the opposite direction.

THINGS TO DO AND THINK ABOUT

1. Study the map showing the locations of tropical storms. Give the names (using an atlas if necessary) of at least six countries affected by:
   (a) hurricanes in the North Atlantic and Caribbean
   (b) typhoons in the Pacific and South China Sea

2. Copy and complete these statements. Choose from the answers listed below.
   storm surge   50 kilometres   1000 kilometres
   252 km/h   27°C   119 km/h
   typhoon   eye
   (a) In a tropical storm, sea surface temperature has to be at least …
   (b) The centre of a tropical storm is called the …
   (c) Sea level rise caused by tropical storms is called a …
   (d) A severe tropical storm in south-east Asia is called a …
   (e) A tropical storm becomes a hurricane when wind speeds reach …
   (f) A category 5 hurricane has wind speeds over …
   (g) The diameter of a tropical storm can be over …
   (h) The width of the calm area at the eye of the storm can be up to …

3. Explain how a tropical storm forms.

 ONLINE TEST

Take the test on tropical storms at www.brightredbooks.net/N5Geography

GLOBAL ISSUES

ENVIRONMENTAL HAZARDS: TROPICAL STORMS – EFFECTS AND MANAGEMENT

It is not possible to prevent tropical storms but with modern weather forecasting techniques and the benefit of geostationary satellites, accurate predictions can be made to give people several days' warning of a storm's approach. This time is invaluable in making preparations to protect property, evacuate people in areas of risk and save lives.

EFFECTS OF TROPICAL STORMS

Hurricanes, typhoons and cyclones have caused some of the most devastating natural disasters on the planet. Their effects are generally more widely felt than those of a volcanic eruption and, because they can last for a week or more and move over areas stretching for thousands of kilometres, the effects are even more widespread than those of a major earthquake.

TYPHOON USAGI

Diagram 3.26 Track of Typhoon Usagi, September 2013: winds up to 296 km/h, category 4 hurricane, pressure at centre 910 millibars, largest typhoon for 30 years, up to 50 people killed, millions affected in the Philippines, Taiwan and China.

Typhoon Usagi formed as a tropical storm on 16 September 2013 in the western Pacific Ocean close to the Philippines and intensified to become a typhoon on 18 September. On the same day, the Joint Typhoon Warning Centre (JTWC) and the Japanese Meteorological Authority (JMA) issued a red alert, with a sustained wind speed of 240 km/h (150 mph) recorded and gusts of 296 km/h (185 mph). Typhoon Usagi had gained the strength of a category 4 hurricane.

At its peak, Usagi measured over 1000 kilometres across and caused torrential downpours of 300–400 millimetres of rain per day over high ground in the Philippines and Taiwan. As the typhoon approached the Luzon Strait between the Philippines and Taiwan, thousands of people were evacuated from their low-lying homes in the northern Philippines provinces on Luzon Island due to the threat of flooding. Many properties were destroyed by mudslides caused by torrential rain, ferries were cancelled, flights were suspended and fishing boats sought shelter in harbours. At least 18 people were killed in floods and mudslides caused by the heavy rains weakening soil on steep slopes. Streets in the port of Subic on Luzon Island were chest deep in water.

In Taiwan 1600 soldiers were deployed to southern areas, which were considered to be at the greatest risk, with 24 000 further troops placed on standby in case of a major rescue operation.

As Usagi tracked across the South China Sea on 21 September, it approached the Chinese coast in Guangdong province. The China State Oceanic Administration issued a category 1 emergency alert, the highest possible level of maritime warning. At least 3.5 million people in China were affected as schools were closed down, ferry services suspended, hundreds of flights in and out of Hong Kong Airport cancelled and at least 80 000 people evacuated from areas at risk of flooding and moved to safety. The port in Hong Kong, one of the world's busiest, was closed to all shipping movements. Hong Kong was threatened by the effects of a storm surge which would raise sea levels at least one metre above the normal highest tide level. Twenty-three temporary shelters were opened to house potential evacuees and victims.

DON'T FORGET

On 8th November, 2013, Typhoon Haiyan made landfall in the central Philippines. This category 5 storm was the strongest ever to hit land with one-minute sustained wind speeds of 315 km/h (196 mph) and gusts of 380 km/h (236 mph). At least 6155 were killed by the typhoon and resulting storm surge. Tacloban, a city of 225 000 people was largely destroyed.

contd

On 22 September, as Usagi made landfall in China's Guangdong Province, wind speeds reached 180 km/h, with trees being toppled and cars blown off the road. In the state capital Guangzhou, all flights were cancelled and high-speed train services suspended. Power supplies were cut off to large areas in both Guangdong and the neighbouring Fujian provinces. During the next two days (23/24 September), Typhoon Usagi dissipated as it moved inland, with its centre some 80 kilometres north of Hong Kong, meaning the city avoided some of the very strongest winds. Nevertheless, this was the most powerful typhoon to hit the area for at least 30 years, causing 25 deaths in China and destroying 7100 homes in Guangdong Province. Economic losses in China alone have been estimated at more than $500 million.

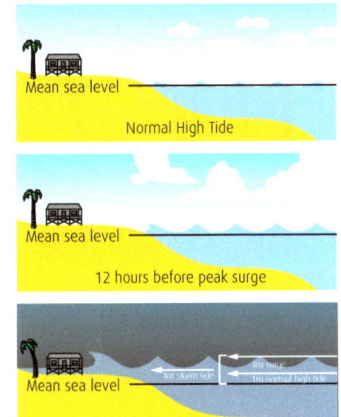

Diagram 3.27 Storm surges are caused by the effects of extreme low pressure on the ocean surface, combined with very strong onshore winds which push the sea towards the land. Intense low pressure results in a higher than usual sea level, while the strength of the onshore winds pushes water towards the coast. This can result in sea levels being raised by 3 metres or more above normal high tide level, swamping low-lying coastal areas and sometimes causing flooding far inland.

LIMITING THE IMPACT OF TROPICAL STORMS

With such a powerful natural force as a hurricane, it may seem that there is little people can do to prevent nature from taking its course. However, modern weather forecasting, aided by satellite images and aircraft which monitor the precise route of the storm, can predict, often with a high degree of accuracy, when and where a storm is likely to hit. It is unusual for people to get as much as a week's warning but often several days' notice is possible.

The key to preventing tropical storms from becoming disasters is to act on the warnings that are received. In the Philippines, the National Disaster Risk and Reduction Management Council can issue warnings and advise people to leave low-lying areas for higher ground. Governments can deploy troops to be ready to help victims in the areas likely to be worst affected by high winds or flooding. Charities such as the Red Cross can stockpile first-aid kits and food packs in preparation for the arrival of a tropical storm, as they did for Hurricane Usagi in Taiwan and the Philippines. Emergency shelters may be opened in advance, as in Hong Kong, with sports halls and schools often being used for this purpose on a temporary basis.

With warnings of the imminent arrival of a storm, people are advised by state authorities to stay at home in order to reduce the number of people who may be out and about and therefore more at risk from, for example, flying debris and falling trees. Schools are often closed and vulnerable transport networks, such as trains, planes and ferries, cancelled until after the storm has passed. In some areas prone to tropical storms, householders may board up windows on properties in the most exposed areas. Rapid response teams of emergency power workers will form one aspect of the plans drawn up by electricity companies in preparation for power outages as a result of broken power lines often due to falling trees disrupting supplies.

In the aftermath of the very worst storms, especially in developing countries with fewer resources, major aid operations will commence, as was the case following Hurricane Tomas in Haiti in 2010. Individual governments, the UN and major charities, such as Oxfam and Medicines sans Frontier, deliver aid as quickly as possible to the worst affected areas. Medical supplies, doctors, nurses, tents, blankets, bottled water and food were all needed following Hurricane Tomas, as Haiti had already been devastated by a major earthquake earlier in the year and was suffering from an outbreak of cholera due to contaminated water supplies.

Effects of a storm surge during Hurricane Sandy in New York City, October 2012.

DON'T FORGET

Hurricanes, cyclones and typhoons can inflict enormous damage on even the wealthiest nations, as with Hurricane Katrina in New Orleans, USA in 2005. However, poorer developing nations have fewer resources to help plan for, or clean up after, a major storm and so it is in developing countries, such as Haiti and Bangladesh, that the effects of tropical storms are often worst.

ONLINE

Check out the link 'Tropical storm risk' to learn more at www.brightredbooks.net/N5Geography

ONLINE TEST

Take the test on tropical storms at www.brightredbooks.net/N5Geography

THINGS TO DO AND THINK ABOUT

1. Write down all the effects on people and the environment of Typhoon Usagi in September 2013.
2. Explain what a storm surge is and how it is caused.
3. What are the main ways in which people can plan for the effects of a severe tropical storm?

GLOBAL ISSUES

TRADE AND GLOBALISATION: WORLD TRADE

International trade is the exchange of goods and services between different countries. The pattern of world trade is complex and countries depend increasingly on each other for resources, services and goods. This is called **interdependence**, where countries rely on each other economically. However, this does not mean that countries are equal when trading. Often there are glaring inequalities that help to perpetuate the difference in wealth between developed and developing countries.

THE PATTERN OF WORLD TRADE

World trade is dominated in terms of total value by relatively few large and powerful nations. The EU, with its 28 member countries, has the largest total share of international trade, although individual countries within the EU, notably Germany, France and the UK, are all very powerful trading nations. The two most powerful individual countries are the USA and China.

Leading world trading nations (2012)

Rank	Country	Total value of goods and services traded internationally (billions US dollars)	Percentage of total international world trade
	World	45 141	100
	EU	5935	13
1	USA	4903	10.8
2	China	4341	9.6
3	Germany	3115	6.9
4	Japan	1998	4.4
5	France	1623	3.6
6	UK	1619	3.6

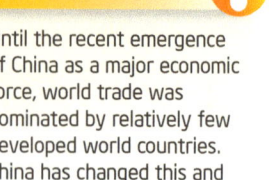

DON'T FORGET

Until the recent emergence of China as a major economic force, world trade was dominated by relatively few developed world countries. China has changed this and has become a major trading partner for many developed and developing world nations.

Exports are the goods and services which a country sells abroad, while **imports** are what a country buys in from other countries. The **balance of trade** is the difference between the value of a country's imports and its exports. A country that sells more than its buys from other countries will have a positive trade balance or **surplus**, while a country which imports more than it exports is said to have a negative trade balance or deficit. In August 2013, for example, the UK had a total **trade deficit** of £3320 million. This was due in part to the large amount of imported consumer goods and a fall in the amount of oil and gas exported from the UK.

It is common for developed countries to import large amounts of raw materials such as iron ore, bauxite, oil, coal and coffee beans. These goods are then processed and sold for a much higher price than was paid to purchase them. For example, iron ore and bauxite are processed and converted into steel and aluminium products, respectively. Some of these can then be exported for a much greater value than the cost of the raw materials. In this way, many developed countries become powerful trading nations, buying in raw materials cheaply and exporting more expensive processed products. Germany and Japan are examples of countries which have done this on a large scale, developing successful economies, although Japan's economy has undergone a period of recession in recent years.

Developing countries tend to export mainly raw materials, such as copper ore and cocoa beans, for which they receive a low price, while they import more processed and manufactured goods, which are much dearer. For this reason, the economies of many developing world countries are in deficit – they have to pay much more for imports than they earn from exports.

Container ships, such as the *RHL Fiducia*, transport much of the world's trade.

contd

Ghana, for example, had a trade deficit of $4559 million in 2012. Its main exports were oil, gold, cocoa beans, timber, bauxite and manganese, while its imports were mainly manufactured goods, petroleum products and foodstuffs. Ghana matches the stereotype of a developing nation with a trade deficit due to the relatively low export value of its primary products (raw materials) and its dependence on expensive imported manufactured goods. However, as its offshore oil industry develops, it is likely that Ghana's economic position will strengthen in the future.

GLOBALISATION

Globalisation is the process whereby countries have become increasingly interconnected and interdependent on each other for the trade of goods and services. Exchange of cultural traditions has also seen global cities becoming more similar in many ways. Companies involved in global trade are now based in many different countries and are said to be multinational. A **multinational company** may still have it headquarters in one country, but will also have bases and factories in many different countries. European aviation giant Airbus, for example, which is part owned by French, German, British and Spanish interests, opened up an aircraft factory in Tianjin in northern China in 2009. By locating an assembly plant here, it is able to access the huge Chinese market more easily while also benefitting from much lower Chinese wage rates. In April 2016, the first aircraft was delivered from the new Airbus factory in Mobile, Alabama, USA.

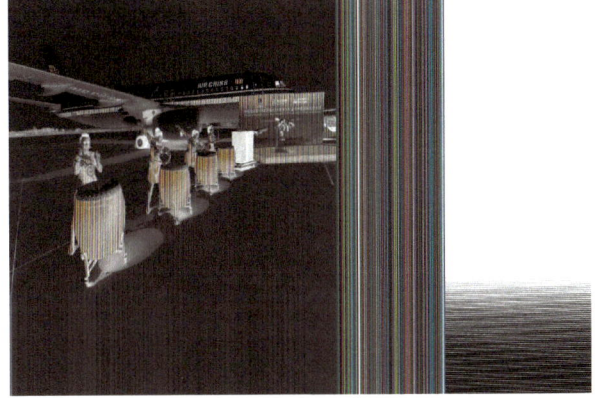

European manufacturer Airbus delivers its 100th Chinese-built A320 aircraft in 2012.

Globalisation has happened because of a huge increase in the amount of goods traded internationally. It has become easier to trade goods because of containerisation involving huge ships which sail continuously via the Panama and Suez Canals around the globe, picking up and dropping off containers holding all sorts of different goods. Air transport has also become easier and cheaper, making it possible to move goods between continents within hours.

New communications networks such as the internet, satellite television and mobile phones have made the exchange of not only goods but also cultural ideas much easier and faster. Groups such as the UN and the World Trade Organisation (WTO) have encouraged countries to trade more easily by the removal of trade barriers, leading to the concept of **free trade**. This is where countries can exchange goods and services without the imposition of import taxes or **tariffs**. The difference in labour costs between the developed and developing world has led to a rapid increase in the amount of manufacturing in countries such as India and China, where multinational companies can set up factories making use of the plentiful and cheap labour supply. This has led to a decrease in manufacturing in developed countries, such as the UK, where wages are higher.

Multinational companies that are examples of globalisation.

Globalisation has had some benefits worldwide but there have also been problems. The next section examines some of these.

THINGS TO DO AND THINK ABOUT

1. Study the multinational company logos. Airbus and Samsung are named. Can you name the other two companies?
2. Make a list of developments which have led to globalisation. You should be able to mention at least five things.

 ONLINE TEST

Take the test on world trade at www.brightredbooks.net/N5Geography

GLOBAL ISSUES

TRADE AND GLOBALISATION: IMPACT OF WORLD TRADE PATTERNS

Globalisation has both positive and negative effects worldwide. Different strategies have been used by countries and organisations to overcome problems created by globalisation. Some countries have set up trade alliances, while fair-trade organisations have also emerged which aim to give producers in developing world countries a better deal.

ADVANTAGES OF GLOBALISATION

The arrival of large multinational corporations has brought benefits to some countries, for example, there are new jobs and workers receive training in new skills. The whole economy of the country concerned can benefit if other companies are attracted in to provide components for, or use the products of, the multinational company. This is sometimes referred to as the **multiplier effect**, where the arrival of one large manufacturer leads to other companies moving into the region. If multinationals buy local resources, goods and services, this helps to bring wealth and foreign currency into the country, which can be reinvested in improving local services such as health and education.

There is also a cultural benefit of globalisation where new ideas, music, food and traditions are introduced from other countries. This, combined with long-established local traditions and culture, helps to enrich communities, bring new experiences and make for more multi-cultural societies. There is also raised awareness of global events and problems, such as overfishing and climate change, which helps to promote the need for environmental protection and sustainable practices worldwide.

DISADVANTAGES OF GLOBALISATION

There are many problems that can result from globalisation. Often multinational companies may exploit workers in developing countries which have weaker labour and environmental legislation. They may be able to pay much lower wage rates and expect workers to work for longer hours with fewer benefits and also have less need to put in place environmental safeguards to stop pollution. If cheaper labour becomes available in another country, multinational companies may close down their factories and move elsewhere, leaving many workers unemployed and damaging the economy. Much of the profit may not stay in the country where goods are being produced, but instead be transferred to the wealthy developed countries where the headquarters of most multinationals are based. Multinational companies may also force local companies out of business as they are able to buy raw materials in bulk and at a cheaper price than small local competitors.

Although globalisation is often seen as benefiting developed countries the most, it has also led to the closure of many factories in the developed world and the redundancy of thousands of workers as companies move abroad to places with cheaper labour. Finally, **cultural globalisation** is seen by many people as a threat to unique cultural identities, languages and traditions, especially of developing world countries. With the advent of satellite TV, the internet and new social media, people may be flooded with ideas, adverts for consumer goods and films, which could lead to the loss of local customs, traditions and even languages. This is sometimes referred to as the loss of cultural diversity.

TRADE ALLIANCES

A **trade alliance** is where two countries, or a group of countries, make an agreement to trade freely with each other. This means that there will be no import taxes or tariffs imposed on goods being exchanged between the countries concerned. Examples of

contd

DON'T FORGET

Sustainability is the idea that activities such as farming and fishing should be carried out in a way that does not harm the environment or limit the resources available for future generations.

Global Issues – Trade and globalisation: Impact of world trade patterns

trade alliances are the EU and the North American Free Trade Association (NAFTA), which comprises Mexico, the USA and Canada. The aim of NAFTA is to eliminate all barriers to trade and investment between the three countries. This gives producers and businesses within North America an advantage over the rest of the world when trading with each other. Free trade allows a wider range of goods and services to be accessed by a country's population and helps to increase the standard of living.

In the EU, the single market allows for the free flow of trade between its 28 member countries and also encourages the free movement of labour within its borders. There are also trade barriers imposed on imports from outside the EU. These include **tariffs** or import taxes and **quotas**, which limit the amount of goods imported from certain countries. The result is that producers within the EU have a trading advantage over those outside the EU. In 17 EU countries, the economic links have been take one step further with the introduction of a common currency, the euro. By having a single currency, business and trade between countries using the euro is cheaper and easier, giving these countries a further trading advantage. However, economic instability within some euro member states has led to a crisis of confidence and an economic downturn within much of the so-called Eurozone.

FAIRTRADE FOUNDATION

Fairtrade are striving for a world where justice and sustainable development are at the heart of trade structures and practices.

FAIR TRADE AND SUSTAINABLE PRACTICES

The fair-trade movement aims to give producers in developing world countries a fairer price for their goods and to help promote sustainable practices. Fair-trade products may be slightly more expensive but this extra money ensures both that exporters in the developing world receive a better price and also that money is put towards environmental and social schemes that help to protect the environment and improve the quality of life for local communities.

Coffee producers who sell their products through the Fairtrade Foundation, for example, have to pay a fee to use the Fairtrade logo on their products and ensure that they meet strict production standards. In return, they receive a guaranteed price for their coffee, even when global prices fall. They also receive the Fairtrade premium, which is money invested in schemes that help to improve living standards, such as building schools and clinics for local people.

These schemes are agreed democratically with the communities concerned. Over 900 Fairtrade products are now sold worldwide, including coffee, tea, bananas, chocolate, cocoa, cotton and sugar. In Uganda, for example, Fairtrade vanilla producers receive at least 8% of the selling price, whereas previously they may have received less than 1%. Extra money is also reinvested in improving the quality of the vanilla and put towards community schemes such as improving water supplies to ensure that they are clean and safe.

Forest Stewardship Council® (FSC) help to promote sustainability.

DON'T FORGET

There are many other organisations which try to ensure sustainable global production practices, such as the Forest Stewardship Council, which promotes responsible management of the world's forests.

Bomarts Farms, Ghana: producer of Fairtrade mangoes and pineapples

- Subsidised canteen meals provided for workers
- Workers receive training in computer skills
- Loans available for all workers
- Two village clinics refurbished
- Day-care facilities provided for workers' children
- Donations towards a maternity wing for the local clinic
- Extra tuition, books and writing materials for workers' children

Diagram 3.28 The benefits of Fairtrade for pineapple and mango producers in Ghana.

ONLINE

Visit the Fairtrade Foundation website to find out more about fair trade and how it helps to improve the quality of life in developing world communities at www.brightredbooks.net/N5Geography

THINGS TO DO AND THINK ABOUT

1. Draw up a table to show the advantages and disadvantages of globalisation. You should aim to find six of each.
2. Explain, in detail, how fair trade helps developing world producers.

ONLINE TEST

Take the test on the impact of world trade patterns at www.brightredbooks.net/N5Geography

109

GLOBAL ISSUES

TOURISM: MASS TOURISM

Mass tourism has become the largest worldwide industry and involves hundreds of millions of tourists travelling to purpose-built resorts, usually in areas of the world where warmth and sunshine are all but guaranteed. In Europe, popular tourist destinations are on the Mediterranean coasts of France, Spain, Italy, Greece, Turkey and Cyprus. However, cheap package holidays may be available even further afield to destinations such as Florida, Cancun in Mexico and even Thailand.

In 2012, international tourist numbers exceeded one billion for the first time. The leading destination was France with over 83 million visitors. Four of the top six destinations were in the Mediterranean.

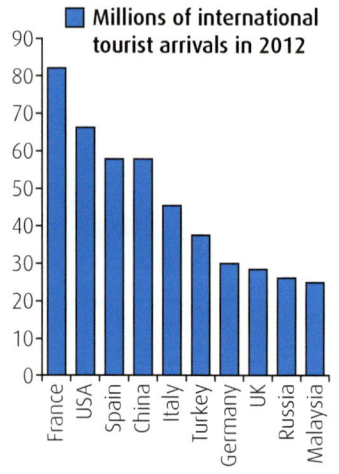

Diagram 3.29 International tourist figures 2012

REASONS FOR THE INCREASE IN MASS TOURISM

Cheap air travel and the promotion of package holidays helped to start off the boom in international tourism in the 1960s. **Package holidays** are where everything required for a holiday is sold together as part of the deal: flights, transfers to hotels, accommodation, the services of a holiday company representative and sometimes even all meals (full board).

People also have, on average, longer holidays now than 50 years ago, as well as more disposable income. This, combined with much improved telecommunications and widespread access to information via the internet, has made it much easier for people to find out about and book foreign holidays. Independent travellers have also boosted tourist numbers. This is where individuals make their own travel and accommodation arrangements directly and without the help of a package holiday company. Often they will visit the same destinations offered by package holiday deals, but sometimes at a lower cost and with more opportunity for flexibility in their holiday arrangements.

A cruise ship passes through Venice.

ADVANTAGES OF MASS TOURISM

For holidaymakers, the mass tourist market has opened up a huge variety of different types of holiday as well as exotic destinations. Travelling to and from holiday destinations is often quick, easy and affordable. Flights to Mediterranean resorts take only between two and four hours and often cost less than flights within the UK. Tourists can indulge in all sorts of different physical outdoor activities, such as walking, diving and water sports and skiing, in climates that are more suited to these activities than in northern Europe. Cultural tourism is also important, where tourists visit areas to sample local traditions, arts and crafts. Often this is combined with cruises where 3000 people or more may visit many different cities over the course of a week or fortnight, while enjoying luxury accommodation on board a large cruise ship.

For countries receiving large numbers of tourists, there are numerous employment opportunities both in the construction of tourist facilities, such as hotels and swimming pools, and in providing services for visitors. This helps to boost local economies and increase the incomes of local people. With higher earnings, local and national governments also benefit by being able to raise more in taxes, which can then be reinvested in local services such as schools and medical care.

The tourist industry requires good local infrastructure, so communities in tourist areas may benefit from the construction of new roads, better bus services and other transport facilities as well as improved broadband connections, gas, electricity and water supplies, and new sewage systems. As tourists from many different nationalities usually visit popular tourist destinations, such as Majorca or the Greek Islands, this can lead to better cultural understanding as both locals and tourists learn about each others' values, traditions, food and music.

Global Issues – Tourism: Mass tourism

DISADVANTAGES OF MASS TOURISM

Employment in tourist resorts can often be poorly paid and seasonal. In winter, with the downturn in tourism, relatively few workers may be kept on. Large international companies may dominate the tourist trade so that much of the profit goes overseas, leaving little benefit for local communities.

Mass tourism has seen the construction of many hotels and other facilities which are far from pleasing to the eye and detract from the local scenery. Often the sheer number of tourists can be overwhelming and it becomes difficult to find any examples of local culture as resorts are swamped by burger bars, British-style pubs, amusement arcades and cheap gift shops, all catering to the mass market. Drunkenness and rowdy and insensitive behaviour can offend local people and lead to cultural tensions. Private beaches established for the exclusive use of holidaymakers can lead to further local resentment.

Tourist hotels require large amounts of land to build on and water supplies for swimming pools, cooking, cleaning and use by guests. This is often unsustainable, leading to the loss of valuable farmland, as well as endangering ecosystems in the areas from which the water is extracted or diverted. Sewage generated by tourist hotels can also lead to the pollution of local rivers and beaches.

- Loss of valuable farmland
- Multiple high-rise hotels can look ugly
- Sewage can pollute the surrounding sea and damage coral reefs
- Private beaches exclude locals from their own coastline
- High rates of water consumption deplete local reserves, damage areas from which it is taken and result in local water shortages

Diagram 3.30 Jamaica: construction of hotels and tourist facilities can have a negative environmental impact.

DON'T FORGET

The economy of a country that is over dependent on tourism can suffer as a result of competition from other holiday destinations. In Jamaica, for example, over 20% of employment is in the tourist industry, making the economy vulnerable when numbers drop.

REDUCING THE IMPACT OF MASS TOURISM IN JAMAICA

There have been tensions between tourists and local people in Jamaica as tourist numbers reached 4.3 million in 2018 on an island of just 2.7 million inhabitants. Often money generated from tourism goes abroad to the multi-national companies running it. Now, across Jamaica, local tourism is being encouraged, whereby tourists stay in smaller, locally-owned hotels within communities rather than in coastal resorts with few local connections. This community-based tourism helps to improve links and cultural understanding between local people and visitors. More profit stays in local businesses instead of going to foreign-owned enterprises.

Hotels have often discharged untreated sewage into the sea off Jamaica's west coast, leading to the loss of more than 30% of coral reefs as sewage encourages the growth of algae, which kill off the coral. In the area around Negril, strict controls have been imposed on hotels to treat sewage waste and an education campaign has made locals and visitors more aware of the importance of coral reefs. Efforts have been made to distribute tourism more widely across Jamaica, away from coastal resorts and into areas such as the Blue Mountains. This has been done by establishing nature reserves and a national park in the Blue Mountains and by recognising tourist interest in the environment. **Ecotourism** is encouraged and small-scale resorts, such as Great Huts near Port Antonio, are making tourism more environmentally friendly and sustainable.

THINGS TO DO AND THINK ABOUT

1. Why has global tourism grown rapidly in recent decades? Try to find at least four reasons.
2. Copy and complete this table. Try to add four or more examples in each column.

Advantages of mass tourism	Disadvantages of mass tourism
Huge variety of different destinations and types of holiday available	Over dependence on tourism can lead to the economy being vulnerable if tourist numbers fall

3. Give four examples of how the negative impacts of mass tourism can be reduced in countries such as Jamaica.

ONLINE

Check out the article 'Six reasons why mass tourism is unsustainable' to learn more at www.brightredbooks.net/N5Geography. What do you think?

ONLINE TEST

Take the test on mass tourism at www.brightredbooks.net/N5Geography

GLOBAL ISSUES

TOURISM: ECOTOURISM

Ecotourism is a type of holiday that is environmentally friendly and sustainable. Usually, relatively small numbers of tourists are involved in order to minimise their environmental impact. Ecotourism holidays often encourage interaction between local people and visitors, who learn about indigenous culture and traditions as well as about the local environment. Destinations can be exotic and remote such as Sarawak in Malaysia, the Amazon rainforest of Peru and the cloud forests of Costa Rica. A key feature of ecotourism is that more of the profit is reinvested into improved services for local community and environmental projects. Ecotourism can therefore be defined as *'responsible travel to natural areas that conserves the environment and improves the well-being of local people'* (International Ecotourism Society).

ECOTOURISM CASE STUDY: COSTA RICA

Diagram 3.31 Protected areas cover 26% of Costa Rica.

Costa Rica is a small Central American country with a population of 4.7 million. It is comparatively well off compared with many other Central and South American countries, has a low birth rate of 1.9 and a high life expectancy of 78. Costa Rica has a rich biodiversity and is well known for its conservation and environmental programmes, with 26% of the country being protected as national parks or nature reserves.

National parks, such as Corcovado, and the other 168 protected areas in Costa Rica bring in an estimated $1.5 billion per year through ecotourism. Many local communities have benefited from ecotourism by finding employment that has improved their standard of living. Previously, the Costa Rican economy relied heavily on agricultural exports, such as bananas, sugar and coffee, so ecotourism has helped bring in much needed foreign currency.

Costa Rica is close to the USA, which accounts for 49% of all visitors to the country. Unfortunately, there has not been enough regulation of what can be classed as ecotourism in Costa Rica, so some companies have been set up that pay little regard to the principles of ecotourism, focusing instead on maximising their profits. This has sometimes led to the sort of problems often associated with mass tourism such as footpath erosion, litter and noise pollution.

Monteverde Cloud Forest Reserve.

MONTEVERDE CLOUD FOREST RESERVE

Monteverde Cloud Forest Reserve lies at an altitude of 1440 metres above sea level in the northern Tilaran Mountains, which form part of the range running down the spine of Costa Rica. The reserve, together with the neighbouring Santa Elena reserve, extends to over 27 500 hectares, with a thriving local population of 7500 people. Monteverde is a good example of a successful ecotourism project. Its altitude ensures a constant supply of moisture, often in the form of mist and cloud, which helps to sustain this diverse forest ecosystem. It is home to over 100 species of mammals, 400 bird species and thousands of different varieties of insects and plants. Visitors can explore this rich and internationally recognised biodiversity from trails and forest canopy walkways.

Hanging walkways allow visitors to experience the forest canopy.

Monteverde is accessible only by a long, winding, bumpy and unsurfaced road. Villagers in the community decided to leave it this way, rather than surfacing the road, in order to keep visitor numbers to manageable levels. The nearest surfaced road is two hours' drive away. Even so, an estimated 250 000 visitors arrive each year to see the cloud forests and their wildlife. In the forests, visitors can see strangler figs, epiphytes and the largest number of orchid species in the world. Among the hundreds of different bird and

contd

Global Issues – Tourism: Ecotourism

animal species to be seen in their natural environment are three-toed sloths, toucans, howler monkeys, tapirs, hummingbirds and jaguars. Visitors are accommodated in eco-lodges or small hotels that make a real attempt to minimise their environmental impact.

There is a network of zip wires and hanging walkways from which visitors have an unparalled view of the forest canopy. On the ground, there are more traditional walking trails, many of which are boardwalks to prevent footpath erosion. Visitors are encouraged to stay on the trails to minimise damage to the forest and to take all litter away with them.

Nature tourism in Monteverde has also led to the establishment of numerous related businesses which provide employment to local people. As well as tour guides, jobs have been created in hotels, restaurants, gift shops, art galleries and cooperative farms which sell produce to visitors through a farmer's market. An organisation called CASEM sells local art and handicrafts to visitors and provides jobs for 150 women from the community, who have seen a rise in their living standards as a result.

Despite the high number of visitors to Monteverde, the environment has been protected and local people have become aware of the ecological importance of the surrounding forest. Whereas it was once under threat from surrounding communities who wanted to clear it to create new farmland, they now see its conservation as being vital to their livelihoods. Local earnings from ecotourism are slightly higher than they were from farming. Villagers and community organisations make the decisions and control which developments are allowed in the area. Up to 80% of the earnings from ecotourism stay in the community and can be reinvested in new services and facilities which benefit both local people and tourists.

Monteverde is an example of an ecotourism project that has protected the environment as well as allowing local people to make a decent living. Visitors and inhabitants both recognise the value of the conservation that underpins Monteverde. Carefully managed tourism has not exploited the environment, has ensured that visitor numbers stay manageable and has been beneficial to the local community. The Monteverde Cloud Forest Reserve is a good example of sustainable ecotourism.

Diagram 3.32 Managing ecotourism in Monteverde.

DON'T FORGET

Income from the entrance fee to the reserve helps fund further environmental protection.

DON'T FORGET

Children from local schools and from further afield in Costa Rica are educated about the forest ecosystem and the importance of protecting it.

ONLINE

Learn more about ecotourism by following the links at www.brightredbooks.net/N5Geography

ONLINE TEST

Take the test on ecotourism at www.brightredbooks.net/N5Geography

THINGS TO DO AND THINK ABOUT

1. What are the attractions of Monteverde Cloud Forest Reserve for tourists?
2. Explain how ecotourism benefits:
   (a) the local community in Monteverde   (b) the environment of Monteverde.

GLOBAL ISSUES

HEALTH: DISTRIBUTION OF DISEASE

INTRODUCTION

Many human diseases are linked to environmental conditions, the standard of living and life style of a community, or any combination of these. Certain diseases, such as **AIDS**, occur worldwide and are found in both developed and developing world countries. Others, such as **cancer**, **asthma** and heart disease, are more common in the developed world, partly because of a longer life expectancy. Diseases such as **cholera**, **malaria**, **pneumonia** and **kwashiorkor** are more prevalent in the developing world. For some of these diseases, there are medicines that offer cures if they are administered quickly enough, but sadly in many cases these medicines are not available because of poverty.

DON'T FORGET

It is a big advantage when you are describing world distributions from a map such as the one in Diagram 3.33 if you can identify individual countries, showing your good geographical knowledge. Study a political world map regularly and get to know the locations of different countries.

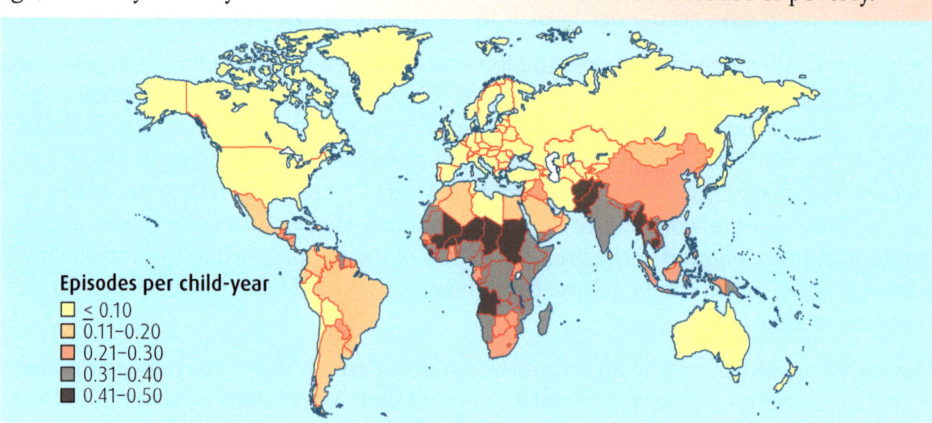

Diagram 3.33 Global distribution of pneumonia (children aged 0–4).

Diagram 3.33 shows the incidence of pneumonia in children under five years old. It is clear that there is a much higher incidence of pneumonia in developing world countries, particularly those in Africa and south-east Asia. In fact, just five countries in the developing world account for over 50% of pneumonia cases in the under-5 age group. These are India, Nigeria, the Democratic Republic of Congo, Ethiopia and Afghanistan. In India alone there are 408 000 deaths of under-5s from pneumonia each year. However, because India has a very large population this represents a lower proportion of its under-5s (0.32%) than, say, Afghanistan, where there are 85 000 deaths of children under five (1.86%). These figures are shocking and indicate high levels of poverty and lack of health care in these areas of the world.

DON'T FORGET

Pneumonia is the leading cause of death in children under five worldwide, with an estimated 1.1 million deaths annually. The majority of these are in the developing world. More children die from pneumonia than AIDS, malaria and tuberculosis combined.

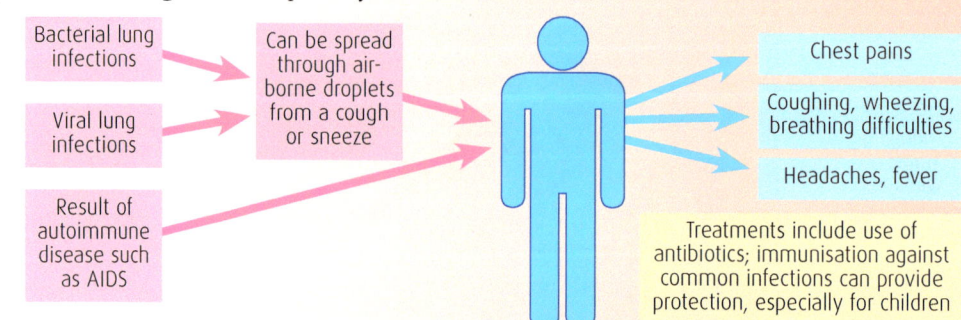

Diagram 3.34 Causes, symptoms and treatment of pneumonia.

CAUSES OF DISEASE

In developing world countries, environmental conditions and low standards of living often explain the high incidence of diseases such as malaria, kwashiorkor, cholera and pneumonia. Infant mortality rates are higher and life expectancy lower than in the wealthier developed world countries. In areas such as South America, Africa and south-east Asia, causes of disease might include:

- lack of safe drinking water and proper sanitation
- malnutrition and famine
- poor standards of housing and hygiene
- lack of education about disease prevention
- shortage of trained medical workers and medicines
- warm, wet conditions which encourage insect pests.

Common to most of these causes are the high levels of poverty in many developing world countries, so even though it should be possible to reduce the incidence of disease, lack of resources makes it very difficult to do so.

Global Issues – Health: Distribution of disease

In wealthier developed world countries, higher standards of living, better sanitation and different environmental conditions mean that diseases, common in the developing world have been largely eliminated. Infant mortality rates are much lower and life expectancy is higher. However, other diseases, such as cancer, heart disease and asthma, are much more common. Causes of these might include:

- unhealthy life styles and lack of exercise, leading to obesity
- poor diets including too much sugar, salt and fatty foods
- hereditary factors can increase the likelihood of cancer and heart disease
- smoking increases the risk of lung cancer, asthma and heart disease
- environmental conditions such as air pollution leading to asthma
- food additives and exposure to chemicals in the environment, which are linked to cancer.

ASTHMA

Asthma is a respiratory condition affecting the upper airways and causes severe breathing difficulties. It is widespread in developed world countries with, for example, three million asthmatics in Japan and four million in Germany. Asthma can also occur in developing world countries and there are estimated to be between 100 and 150 million people who suffer from asthma globally, with approximately 180000 deaths annually.

There is no single cause of asthma, but a number of things can increase the likelihood of a person developing it:

- a family history of asthma or other related allergic conditions such as eczema
- exposure to pollen, house dust mites or mould spores due to damp housing conditions
- allergic reactions to certain types of food, animal fur or feathers
- airborne irritants, such as cigarette smoke, or atmospheric pollution such as exhaust fumes
- chest infections such as colds or the flu, caused by viruses
- sudden changes in weather, such as to very cold air or very warm humid conditions, can trigger an asthma attack.

Asthma inhalers for relief and prevention of attacks.

The effects of an asthma attack can be sudden and very serious. There is a narrowing of the upper airway, which can be caused by inflammation or a tightening of the muscles around it. This may cause coughing, wheezing, severe breathing difficulties and a racing pulse. Often mucus or phlegm narrow the passages of the airways even further and compound breathing problems. The threat of an attack can limit the physical activities that an asthmatic person is able to do as vigorous exercise can trigger one.

Treatment of asthma aims to get a person's breathing under control quickly. Often this is done by using an inhaler, which allows medicines to be delivered directly into the airways when breathing in. There are two main types of inhaler. Reliever inhalers are designed to stop breathing problems quickly by relaxing the muscles around the narrowed airways, allowing regular breathing again. Preventer inhalers work over a longer period of time and aim to reduce the inflammation that can lead to an asthma attack. There are also medicines such as steroid tablets that can be taken by asthmatics for whom inhalers are not providing sufficient relief. Usually these medicines are only taken over a short period of time because of their possible side effects.

ONLINE

Check out the links 'Pneumonia' and 'Asthma' to learn more at www.brightredbooks.net/N5Geography

ONLINE TEST

Take the test on world distribution of disease at www.brightredbooks.net/N5Geography

Other ways of reducing the likelihood of asthma involve targeting the environment in which an asthmatic person lives. Ensuring that houses are not damp or mouldy reduces the likelihood of breathing in mould spores. An allergic reaction to house dust mites can also lead to an asthma attack so making conditions less favourable for them helps to reduce the risk. This can be done by keeping a house well aired (reducing dampness), regular cleaning and vacuuming, and the use of barrier covers on mattresses and bedding.

THINGS TO DO AND THINK ABOUT

1. (a) Describe, in detail, the world distribution of pneumonia cases in children under five?
   (b) Give reasons why these parts of the world have such high rates of pneumonia.

2. Make two spider diagrams, one to show the effects of asthma and one to show possible methods of treatment.

GLOBAL ISSUES

HEALTH: DEVELOPED WORLD DISEASES

Asthma, AIDS, cancer and heart disease are all illnesses that are prevalent in developed world countries such as the UK. Asthma was described on page 115 and AIDS is covered later in this chapter. Cancer and heart disease are the two biggest causes of death in Britain, accounting for over 55% of deaths in 2018.

CANCER: CAUSES AND EFFECTS

There are over 200 different types of cancer and it is not clear how all of these are caused. However, there are a number of clearly identified risk factors that make it more likely that a person will develop the disease:

- *Age:* 63% of people who get cancer are aged 65 or over.
- *Lifestyle:* Our lifestyle choices increase the risk of certain cancers. For example prolonged exposure to sunlight can increase the risk of skin cancer, smoking increases the risk of lung cancer and other types of cancer, heavy alcohol consumption increases the risk of liver cancer.
- *Diet:* Unhealthy eating, including the consumption of too much red meat and processed meats and a lack of fresh fruit and vegetables, can increase the risk of breast, bowel and prostate cancers.
- *Obesity* has been linked with increased cancer risk.
- *Inherited risk:* A family history of certain cancer types significantly increases the risk of a person contracting them.

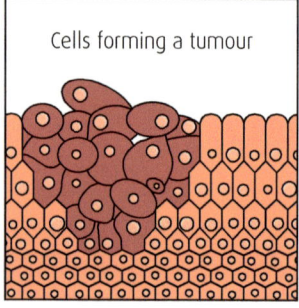

Diagram 3.35 How a tumour forms.

Cancer is a disease that affects the cells in different parts of the body. Normally, they divide and reproduce themselves in an orderly and controlled way. For some reason in cancer this process is disrupted and cells multiply out of control, forming a lump or **tumour**.

Sometimes a tumour is said to be benign, meaning it doesn't spread to other parts of the body, but if it is malignant it can spread rapidly, often via the body's lymph node system. Apart from lumps, other symptoms of cancer include unexplained bleeding, such as coughing up blood, sudden weight loss that cannot be explained by exercise, fatigue and changes to the skin such as to the shape of an existing mole.

Cancer is a life-threatening disease that can progress very quickly but in many cases it can also be treated successfully if it is diagnosed early enough.

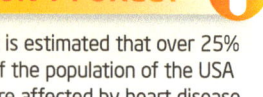

DON'T FORGET

It is estimated that over 25% of the population of the USA are affected by heart disease. As many as 2400 Americans die from it every day.

HEART DISEASE: CAUSES AND EFFECTS

Coronary heart disease is usually the result of fatty deposits that build up on the walls of arteries close to the heart and impede the flow of blood. There are a number of factors strongly linked to heart disease that increase a person's risk of contracting it:

- smoking
- high blood pressure (also referred to as hypertension)
- lack of exercise
- diabetes
- family history of heart disease
- poor diet
- obesity or being overweight.

Often there may not be any single cause of heart disease, but a combination of some of these factors may lead to a person suffering from the condition. People in developed world countries are much more likely to have a sedentary but stressful lifestyle and an unhealthy diet, both of which increase the risk of heart disease.

The effects of heart disease may build up unnoticed over a long period of time. However, if fatty deposits, such as cholesterol, build up on the walls of an artery, it can become hardened or narrowed, leading to **angina** or chest pains because the heart is getting insufficient oxygen to work properly. This can lead to the heart becoming weakened and eventually to heart failure. Other symptoms of heart disease include breathlessness, irregular heartbeat, fatigue and dizziness.

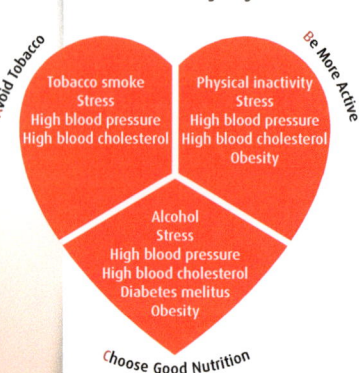

Diagram 3.36 Heart disease.

Global Issues – Health: Developed world diseases

STRATEGIES USED TO MANAGE CANCER

The treatment of cancer varies according to the type of cancer and the stage the disease has reached. Treatment can involve surgery to remove a tumour or the use of radiation to target and eliminate, or at least reduce the size of, a tumour. These are **local therapies**, in that they target a tumour in one part of the body. **Systemic treatments** involve the use of drugs (**chemotherapy**) that target the cancer throughout the body via the bloodstream. This helps to kill off, or at least slow the growth of, cancer cells that have spread beyond the original tumour.

There is no single cure for cancer and many of the treatments used can have severe side effects such as fatigue and, with chemotherapy, lowered immunity to infections and temporary hair loss. However, many courses of treatment are successful and cancer patients can go on to lead full and active lives.

Health education is a major way of reducing the incidence of cancer by teaching people about lifestyle changes that they can make to reduce their chances of developing the disease. These include stopping smoking, eating a balanced and healthy diet (e.g. eating five portions of fruit and vegetables each day), taking regular exercise and keeping to a healthy weight.

Diagram 3.37 NHS five-a-day healthy eating campaign.

STRATEGIES USED TO MANAGE HEART DISEASE

Successful management of heart disease in the UK has helped to reduce deaths from this condition by 40% in the under-75 age group since 2000.

As with cancer, management of heart disease involves treating the disease itself and also educating people about lifestyle choices that will help to reduce their chances of suffering from it.

Treatment for patients with heart disease includes the use of various medicines. Some medicines are designed to thin the blood and make it less likely to clot (e.g. aspirin), while others will help to reduce cholesterol levels (e.g. statins) or treat the effects of angina and high blood pressure. Often surgery is required to help open up and widen narrowed or blocked arteries. This can involve the use of a **stent** or tiny balloon that will inflate and widen the blood vessel (this procedure is called **angioplasty**). In severe cases, a **heart bypass** operation or even a **heart transplant** may be necessary.

Health education programmes designed to reduce levels of heart disease have seen some success in many developed world countries. As people have become aware of the health risks associated with heart disease, their motivation to make lifestyle changes has increased. Simple measures such as taking regular exercise, not smoking, eating healthily and reducing intake of fatty foods, as well as keeping to a healthy weight, can all help to reduce the risk of heart disease. However, heart disease is still a major problem in the developed world and the reduction in the number of cases has not benefited every level of society. Often it is people who are wealthier and better educated who have been able to make the lifestyle changes that have brought about a reduction in cases of heart disease. Other sections of society have not seen the same reductions and so there is still a need for more health education.

 ONLINE

Check out the links 'Cancer' and 'Heart disease' to learn more at www.brightredbooks.net/N5Geography

 ## THINGS TO DO AND THINK ABOUT

1 Copy and complete the table below. Try to add four more statements in each column.

Treatment of cancer	Health education strategies to reduce cancer risk
1. Use of radiation to target tumours	1. Not smoking

2 Construct a similar table for heart disease.

 ONLINE TEST

Take the test on developed world diseases at www.brightredbooks.net/N5Geography

117

GLOBAL ISSUES

HEALTH: AIDS – A GLOBAL DISEASE

AIDS (acquired immunodeficiency syndrome) is a disease that occurs worldwide and can be fatal. It is, therefore, a pandemic disease and has been responsible for the deaths of over 30 million people since it was first recognised in 1981. AIDS is spread by the **human immunodeficiency virus (HIV)**, which can survive in a person's body for many years before any symptoms become apparent. The virus attacks the body's immune system so that it can no longer protect itself from infections. AIDS is the final stage of HIV infection where the body cannot fight off life-threatening illnesses. In 2020, UNAids (the UN organisation leading the fight against AIDS) estimated that 680 000 people die from the disease annually worldwide and that an estimated 37.7 million people are living with AIDS. There is no cure for AIDS but most people with HIV can be treated so that they can live a long and healthy life and are unlikely to go on to develop full-blown AIDS.

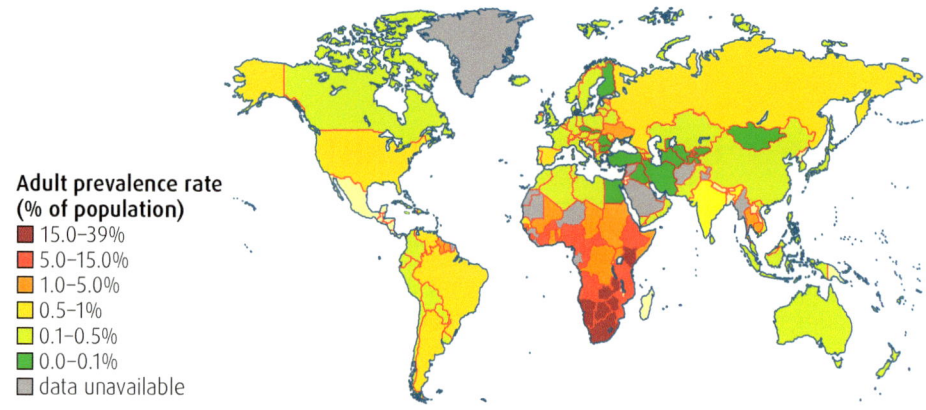

Diagram 3.38 Percentage of adult population which is HIV positive.

CAUSES AND EFFECTS OF AIDS

DON'T FORGET

One of the biggest reasons for the rapid spread of AIDS, particularly in the developing world, was lack of awareness of how the disease spreads and, until recently, the absence of effective advertising campaigns about how to reduce the risks.

AIDS is caused by HIV, a virus that cannot live for long outside the body but is most commonly transmitted from person to person via infected bodily fluids. This usually happens during sex without the use of a condom. It can also be passed on by needle sharing among drug users and from an HIV-positive mother to her child during pregnancy and breastfeeding. It cannot be spread by spitting or perspiration. In the UK, the Health Protection Agency estimated that 95% of HIV cases are contracted as a result of unprotected sex. In the developed world, blood used in transfusions is screened for HIV and so it cannot be transmitted this way, but in some developing world countries, particularly in Africa where screening is not as efficient, there is a risk of HIV infection from blood transfusions and also from infected implements used in surgery. Up to 15% of HIV infections could result from contaminated blood transfusions and surgery in some developing world countries.

DON'T FORGET

HIV is a viral infection which may eventually lead to the disease known as AIDS.

People living with HIV are often unaware that they are infected as the disease can take a long time to show any symptoms. Initially, HIV may cause flu-like symptoms that can last for up to 4 weeks, but the same symptoms can be caused by other conditions too and so are not necessarily evidence of HIV. A simple saliva test can be done to determine whether or not a person is infected by HIV, and this is recommended where a person has put themselves at risk of contracting the disease.

After the initial effects, it may be several years before any other signs of AIDS appear. These are referred to as late-stage HIV infection and may include:

- persistent tiredness and night sweats
- weight loss
- blurred vision
- white spots on the tongue or mouth
- dry cough and shortness of breath
- fever of above 37°C lasting several weeks
- persistently swollen glands.

contd

Global Issues – Health: AIDS – a global disease

As a result of a severely suppressed immune system, a person with late-stage HIV is more prone to life-threatening diseases such as tuberculosis, pneumonia or cancer. While there is no cure or vaccine for AIDS, patients in developed world countries, such as the UK, can be given drug treatments that will control some of these infections and reduce the rate of the spread of other symptoms. However, in developing world countries where there is a shortage of health care and drugs, the death rate from the disease is much higher. In addition, the disease is more likely to go undiagnosed for much longer and people suffering from AIDS, but unable to receive treatment, become increasingly weak and too tired to work, causing severe economic problems for their families.

World AIDS day logo.

MANAGEMENT OF AIDS

There is no cure or vaccine to treat HIV infection, but if anti-HIV medication is taken within 72 hours of likely exposure to the virus it can sometimes be possible to prevent a person from becoming infected. However, once an HIV test has proved positive, treatment can be started using antiretroviral drugs (ARVs). Usually, a combination of ARVs is used to stop the virus from quickly adapting and becoming immune to a single medicine. This approach is often referred to as combination therapy or antiretroviral therapy (**ART**). Patients treated with ART before the onset of late-stage HIV can often go on to lead full and active lives.

There is a wide discrepancy between the developed and developing world in the availability of drugs for the treatment of AIDS. Some developing world countries lack the resources to fund extensive treatment of AIDS and there has been a lack of HIV awareness, which has led to AIDS becoming much more widespread in countries such as South Africa. Whereas health education about HIV and AIDS is well established in developed world countries, often this was lacking in the developing world. Campaigns, such as World AIDS Day, which has taken place on 1 December every year since 1995, have been developed to raise global awareness of the disease.

HIV infection rates in selected countries.

Country	Percentage of population HIV positive	Number of people HIV positive
Swaziland	26	0.2 million
Lesotho	23	0.3 million
Botswana	23	0.3 million
South Africa	17	5.6 million
Zimbabwe	15	1.2 million
Zambia	13	1.0 million
Namibia	13	0.2 million
USA	0.6	1.1 million
India	0.3	2.4 million
UK	0.3	0.1 million
World	0.8	35.3 million (2012)

In recent years, there have been some successes in the battle against AIDS, particularly in the developing world. Global deaths from the disease have been reduced by 42% from the peak in 2004 to 1.2 million in 2014. Annual cases of HIV infection have also fallen 35% since 2000 to 2.0 million in 2014, with over 30 countries showing a 50% decrease in infection rates. This has been due to increased awareness of HIV and AIDS, especially in developing world countries. The price of treatment has also been reduced dramatically, with the cost of ARVs falling from £6200 per person in the mid-1990s to just £87 in 2013, making it easier for people in poorer countries to receive treatment. With raised AIDS awareness, more pregnant HIV-positive mothers have been accessing services such as taking ARVs, to prevent the risk of transmission to their baby.

ONLINE

For more information on AIDS, go to the World AIDS Day link at www.brightredbooks.net/N5Geography

AIDS

- Use of and reducing the cost of antiretroviral drugs (ARVs)
- Sexual health education
- Improved distribution of and access to condoms
- Improved screening of blood used in transfusions
- Fund raising and awareness raising, e.g. World Aids Day
- Antenatal treatment of HIV-infected mothers, preventing transmission to unborn babies
- Advertising campaigns to remove AIDS-associated discrimination
- Raising awareness amongst drug users: needle exchange schemes
- Continuing research into new medicines and vaccines

Diagram 3.39 Management of HIV and AIDS.

THINGS TO DO AND THINK ABOUT

1. HIV infection can lead to AIDS. Give three ways in which HIV can be transmitted.
2. What are the effects of AIDS on a person who contracts HIV?

ONLINE TEST

Take the test on AIDS at www.brightredbooks.net/N5Geography

GLOBAL ISSUES

HEALTH: MALARIA – A DEVELOPING WORLD DISEASE

Malaria is a parasitic, life-threatening disease transmitted by the female anopheles mosquito. It is found in more than 100 countries, mainly those with warm, humid, tropical climates. The World Health Organisation estimates that at least 409 000 people died from malaria in 2019, with up to 229 million cases of the disease worldwide. Some estimates put the number of malaria deaths annually as high as 1 million. People living in the poorest countries are most at risk of catching malaria, with 90% of worldwide deaths occurring in sub-Saharan Africa. Just two countries, Nigeria and the Democratic Republic of Congo, account for 40% of all deaths from the disease.

ONLINE

For more information on malaria check out the link to the UN Roll Back Malaria campaign at www.brightredbooks.net/N5Geography

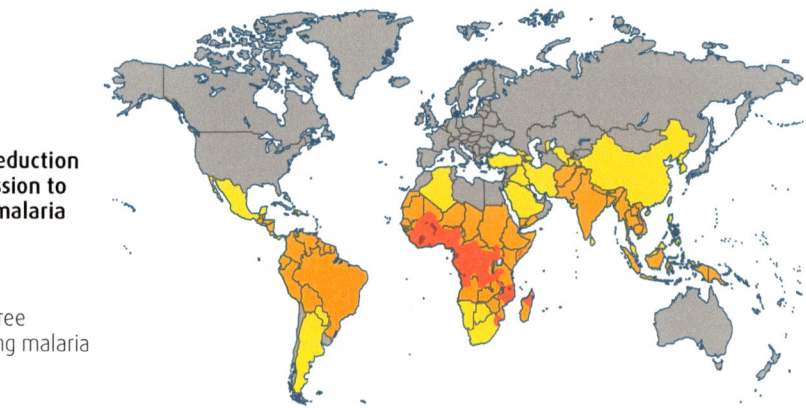

Diagram 3.40 Map showing areas affected by malaria.

DON'T FORGET

3.3 billion people, almost half the world's population, are at risk from malaria. Awareness and fund-raising campaigns such as World Malaria Day, held each year on 25 April, aim to provide solutions in the fight against the disease.

MALARIA: CAUSES AND EFFECTS

Malaria is an **endemic disease**. This means that it is found only in certain populations or locations. The **anopheles mosquito** is responsible for spreading (although not directly causing) the disease. It can live in a wide variety of different climates, but in many parts of the world the disease has been successfully eliminated, so mosquitos cannot transmit it. However, in most tropical areas malaria parasites exist within the population and can be transmitted from person to person by this mosquito.

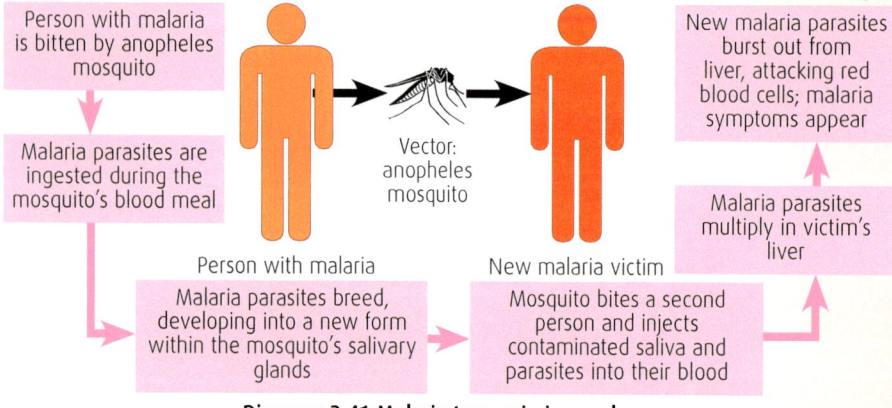

Diagram 3.41 Malaria transmission cycle.

Malaria is caused by a parasite that lives and multiplies within the human body. When a person with malaria is bitten by an anopheles mosquito, malaria parasites are sucked into its gut during a blood meal. These parasites breed in the gut of the mosquito and then migrate to its salivary glands. When the mosquito bites another person, it pumps saliva containing malaria parasites into their blood to stop it clotting, whilst taking another blood meal. The parasites then migrate to the liver of their new victim where they multiply before breaking out in a new form to attack the red blood cells. This is when malaria symptoms start to appear, approximately 2 weeks after the person was bitten.

Symptoms may include fever, headaches, sweating, chills, vomiting and muscle pains. In its most serious form (**falciparum malaria**), these symptoms can rapidly become life-threatening if treatment is not started quickly. It is often young children who are most vulnerable, as they have not yet built up natural resistance to the disease.

contd

Global Issues – Health: Malaria – a developing world disease

Malaria can reoccur and people can go through several bouts of illness. This can have a very serious economic effect, both on families whose main breadwinner cannot go out to work and on countries which consequently have a severely depleted workforce. Diseases such as malaria therefore perpetuate the cycle of poverty in which many communities find themselves. Poor people cannot protect themselves or afford treatment for malaria and so become ill. They cannot go to work, lose income and their family becomes poorer. Malaria can severely weaken the body's defences against other diseases and so a person suffering from malaria is more likely to fall victim to other illnesses. In some rural parts of Africa, up to 40% of crops may go unharvested as workers are too weak or ill to work in the fields due to malaria.

Cost of antimalarial treatment.

Type of treatment	Cost
Rapid diagnostic test for malaria	£0.33
ACT pills to treat an adult for uncomplicated malaria in 3 days	£0.63
ACT pills for 3-day course of treatment for an older child	£0.44
ACT pills for 3-day course of treatment for children under 5	£0.28
Treatment by injection for child with severe malaria	£1.88

(Source: adapted from Medicines sans Frontieres)

MANAGEMENT OF MALARIA

Malaria can be tackled using medicines for people already affected by it, or by using a variety of prevention methods designed to reduce the spread of the disease.

The malaria parasite has a history of quickly adapting and becoming **resistant** to medicines designed to kill it. Drugs containing **quinine** have been extensively used in the past but the malaria parasite has become resistant to these in many parts of the world. Related drugs, such as **chloroquine** and **mefloquine**, can be used both to treat malaria and to prevent people from contracting it.

Recent developments have seen the introduction of **artemisinin**-based drugs derived from the **artemisia** plant. In order to reduce the chances of malaria parasites becoming resistant to this new type of medicine, **artemisinin-based combination therapy (ACT)** has been widely used as a treatment for malaria.

The cost of antimalarial drugs (such as ACT pills) has often reduced access to effective treatment in the developing world, but costs have fallen dramatically in recent years, making these drugs much more affordable, as shown in the table. Unfortunately, extreme poverty and remote locations still leave many parts of the developing world with little or no effective access to treatment for malaria.

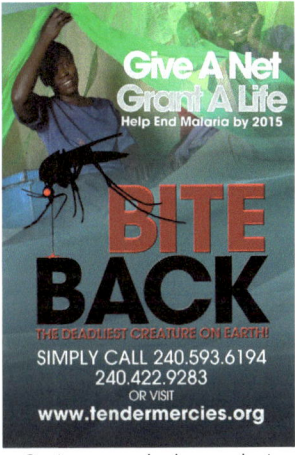

Poster campaigning against malaria.

Many different measures have been used to prevent the spread of malaria. These aim both to reduce the population of the anopheles mosquito in the environment and to educate people about ways in which they can reduce their chances of being bitten. Strategies used have included:

- spraying insecticides on the insides and outsides of houses to kill mosquitoes (**DDT** used initially but as mosquitoes mutated and became resistant, other insecticides such as **malathion** were developed)
- eradicating the breeding sites of the anopheles mosquito (often by draining marshy land or flushing out areas of standing water, as the mosquito larvae take 7 days to develop in stagnant water)
- planting mustard seeds in areas of water which crowd out and eventually drown mosquito larvae
- spreading egg white or cooking oil on the water surface to drown larvae
- use of fermenting coconuts containing bacteria which kill mosquito larvae but leave other organisms in ponds unharmed
- fish introduced into rice fields to eat the mosquito larvae
- use of bed nets soaked in insecticide to prevent people being bitten whilst asleep (each net costs about £3.30 but lasts for at least 3 years)
- education including songs to teach people about malaria prevention (useful in areas of high illiteracy)
- avoiding going out at dawn and dusk when mosquitoes are most active
- wearing long clothing to reduce areas of exposed skin
- use of **DEET**-based insect repellents.

The fight against malaria today continues to focus on the development of an effective vaccine as well as new drugs to combat the continually evolving strains of malaria parasite as they become resistant to each new medicine.

THINGS TO DO AND THINK ABOUT

1. Explain how malaria is caused.
2. Describe the effects of malaria on individuals and communities in the poorest areas of the developing world.
3. Study the map in Diagram 3.40 and describe, in detail, the global distribution of malaria.

ONLINE TEST

Take the test on malaria at www.brightredbooks.net/N5Geography

GLOBAL ISSUES

HEALTH: CHOLERA AND KWASHIORKOR – DEVELOPING WORLD DISEASES

Both of these diseases are easily preventable and occur mainly in the poorest areas of developing world countries where poverty results in either insanitary living conditions, which might lead to the spread of **cholera**, or poor diet, which can result in **kwashiorkor**. Cholera outbreaks can also occur following major disasters such as the 2010 Haiti earthquake, where the existing water and sewerage infrastructure was destroyed, leaving people vulnerable to infection due to contaminated food or water.

DON'T FORGET

Cholera is a deadly disease that affects the most vulnerable communities such as refugees and those living in extreme poverty. It can be fatal, even for a healthy adult, within hours.

CHOLERA: CAUSES AND EFFECTS

Cholera is a bacterial infection that is caused by people consuming water or food that is contaminated with the *vibrio cholerae* bacteria. These are usually passed on when clean water supplies are polluted by sewage, or simply due to poor hygiene when people are handling food.

Although as many as 75% of people who are exposed to the *vibrio cholerae* bacteria do not develop symptoms, they become carriers and can easily pass the disease on to others. The main symptoms of cholera appear within 2–5 days of infection and are severe diarrhoea and vomiting, nausea and muscle cramps. The main danger from cholera is that a person can quickly become dehydrated due to loss of body fluids. This can lead to shock and a severe drop in blood pressure, and can be fatal if not treated quickly.

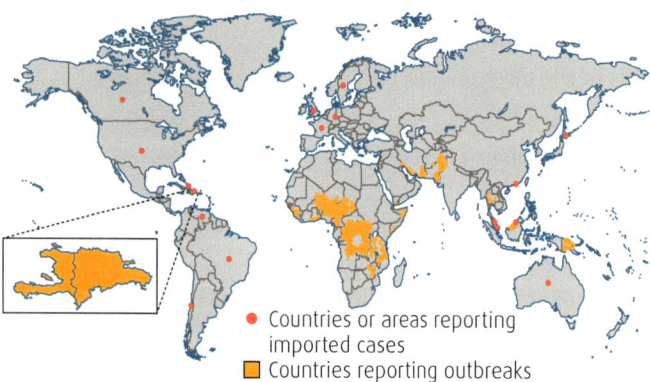

- Countries or areas reporting imported cases
- Countries reporting outbreaks

Diagram 3.42 Cholera outbreaks 2011–2012.

Cholera outbreaks can have a devastating effect on the poorest communities, where poor diet or exposure to other diseases can leave a person's immune system in a weakened state and more vulnerable to attack from cholera bacteria. Where lots of people are living close together in insanitary conditions, as a result of natural disasters or war, the threat of a cholera outbreak is very high. Children are particularly at risk and if not treated they can die within 24–48 hours. Without the right treatment, over 60% of people admitted to hospital with cholera will die, but if it is treated quickly and correctly this is reduced to a mortality rate of under 1%.

STRATEGIES USED TO MANAGE CHOLERA

For someone with severe symptoms of cholera, the immediate replacement of lost body fluids is the priority. This can be done using **oral rehydration solution (ORS)**. A sachet containing vital minerals, salt and glucose is mixed with clean water to form a drink. This will help to stop dehydration and replace vital body salts. At the same time, antibiotics such as **doxycycline** can be administered to treat the underlying infection. For people travelling to areas where there is a cholera risk, such as a disaster zone, there is a vaccination (given as a drink) that is 85% effective.

To prevent cholera outbreaks, the provision of safe water supplies and proper sanitation systems is vital, as is education about basic hygiene. These are always priorities for aid agencies following disasters such Typhoon Haiyan in the Philippines (November 2013). Chlorine can be used to sterilise water, and soap provided so that people can prevent the spread of the disease. Cholera can also be prevented by ensuring that water is boiled and that food is served piping hot. Ice cubes, raw fruit and vegetables, such as salad, could be sources of cholera bacteria if they have not been prepared hygienically.

Oral rehydration solution sachet: in 2012, UNICEF supplied 71 million sachets.

contd

Global Issues – Health: Cholera and Kwashiorkor – developing world diseases

Cholera is a particularly dangerous disease in that it can spread quickly and can be fatal within 24 hours. Yet it can be easily prevented through a combination of preventative measures, use of antibiotics and ORS sachets which cost just a few pence each.

KWASHIORKOR: CAUSES AND EFFECTS

Kwashiorkor is a severe form of malnutrition caused by lack of protein in the diet. It is most commonly found in children in poor communities in tropical or sub-tropical countries of the developing world. It frequently happens when an infant who has been breast-fed (breast milk is rich in proteins) is weaned after about a year, when the next baby is born, and has to survive on a diet of white rice, yams or cassava. These are foods which are rich in carbohydrates but which contain almost no protein. As a result the child develops kwashiorkor because their impoverished family is too poor to give them sufficiently nourishing food.

The symptoms of kwashiorkor include puffy skin caused by oedema (excess water retention in body tissues), which can lead to the abdomen swelling, giving the appearance of a pot belly. Other symptoms are tiredness, an enlarged liver, loss of weight and muscle mass, inflamed red and flaking skin, stunted growth and hair which becomes brittle and loses its colour. Children with kwashiorkor are also likely to have a weakened immune system and therefore become susceptible to other infections and diseases. Without treatment, kwashiorkor can be fatal.

VIDEO LINK

Have a look at the 'Nutritional Emergency in the Southwest' video at www.brightredbooks.net/N5Geography to see the effects of kwashiorkor.

STRATEGIES USED TO MANAGE KWASHIORKOR

Kwashiorkor can be treated by first stabilising levels of body fluids and treating any infections. As levels of salt and minerals in the bloodstream might be quite low, vitamin and mineral supplements are administered. Zinc supplements can help the skin to recover. After skin swelling has been reduced, within about 48 hours, small amounts of food can be reintroduced. Firstly carbohydrates to give energy and then foods which are high in protein such as milk and dairy products, fish, meat, beans, nuts and seeds.

Around 85% of children treated for kwashiorkor will make a full recovery but if treatment is delayed and malnutrition is advanced, a child's growth may be stunted and they may also suffer from physical and intellectual disabilities.

The key factor in managing kwashiorkor is prevention. As kwashiorkor is a diet-deficiency disease, it can be prevented by eating a well-balanced diet. The focus of many aid agencies is on teaching the most impoverished communities about the need for a balanced diet and how they can grow the right type of foods that will increase their protein intake. This might include planting crops such as peanuts, cashews, lentils, sunflowers or soya beans. Improving crop yields by, for example, providing irrigation schemes in areas of drought can also help to make the right types of food available. Another strategy in the battle to control kwashiorkor includes education about the importance of family planning in order to limit family size and therefore increase the availability of food for each child.

Protein-rich crops can be grown to prevent kwashiorkor: sunflower field in Pakistan.

DON'T FORGET

Cash crops for export are often a vital source of income in developing world communities and so people are often left without enough food themselves, leading to diseases such as kwashiorkor.

THINGS TO DO AND THINK ABOUT

1. Study the world map and describe, in detail, the distribution of cholera outbreaks in 2011–12.
2. (a) Explain why cholera outbreaks are often a threat following major disasters such as earthquakes or typhoons.
   (b) How can this threat be reduced in disaster situations?
3. What is oral rehydration solution and how can it be used to treat cholera?
4. Explain how kwashiorkor can:
   (a) be treated
   (b) be prevented.

ONLINE TEST

Take the test on cholera and kwashiorkor at www.brightredbooks.net/N5Geography

GLOSSARY

abrasion
erosion caused by scraping and scouring action of rock fragments carried by glaciers (or rivers, waves)

accessibility
how easy a place is to get to

ACT
artemisinin-based combination therapy: the treatment of malaria patients using a variety of medicines, including artemisinin, to reduce the likelihood of malaria parasite resistance

afforestation
planting trees

AIDS
acquired immunodeficiency syndrome: a disease caused by the human immunodeficiency virus (HIV)

air mass
a large body of air with specific characteristics which affects the weather when it arrives in an area

alluvium
fertile material left behind after a river floods (also called silt)

alternative energy
renewable energy sources such as wind, wave or solar power that are clean and do not pollute the environment in the way that fossil fuels do

altitude
the height of a place above sea level

anemometer
an instrument that records wind speed

angina
chest pains due to obstruction of the coronary arteries

anticyclone
an area of high pressure (usually brings fine, dry weather)

ANWR
Arctic National Wildlife Refuge

appropriate technology
small-scale sustainable developments, often using local resources, designed to improve living conditions in the developing world

aquaponics
where fish farming is combined with hydroponics

arable farming
growing crops

Arctic maritime
air mass originating in the Arctic bringing very cold weather and snow in winter

arête
a narrow, knife-edged ridge, formed by glacial erosion

ART
antiretroviral therapy: the treatment of HIV-positive patients with antiretroviral medicines (ARVs)

artemisinin
medicines used to fight malaria derived from the artemisia plant

aspect
the direction which a slope or settlement faces

asthma
respiratory condition affecting the upper airways, often causing severe breathing difficulties

attrition
where stones and pebbles are gradually made smaller by erosion

backwash
water moving back down a beach due to gravity after a wave has broken

balance of trade
the difference between the value of a country's imports and the value of its exports

barrios
shanty town in Spanish-speaking Central and South America

bay
a wide curving coastal inlet

bedding planes
the boundary between different layers of sedimentary rock

biofuels
fuel produced by converting living organisms, such as plants or algae, into energy

biogas
fuel made from fermenting animal waste

birth rate
the number of live births per 1000 people per year

blowhole
a cave where part of the roof is eroded all the way to the surface, sometimes causing spray to be ejected from it in stormy weather

boulder clay
a mixture of glacial moraine, rocks and rock fragments

brownfield site
land in an urban area that has been built on before and is awaiting redevelopment

bustee
shanty town in India

buttress roots
large spreading root systems needed to support the great height and weight of trees in the tropical rainforest

cancer
a disease causing cells in the body to multiply out of control, forming a lump or tumour

canopy
dense tree foliage high up off the ground blocking out sunlight to the areas below

CAP
Common Agricultural Policy (see below)

carbonation
the process by which limestone rock is turned into a soluble substance on contact with mildly acidic rainwater

carbon footprint
impact which something has on the atmosphere through the production of greenhouse gases

Carboniferous limestone
a sedimentary rock made from shells of tiny sea creatures which accumulated on the sea bed during the Carboniferous geological era

cavern
a large cave found in an area of limestone

catchment area
the land from which a river and its tributaries gather their water; also referred to as the drainage basin

CBD
central business district (see below)

census
national population survey carried out to gather information about the population in a country

central business district
the commercial and business centre of a town where land values are highest

charitable aid
aid given by charities, where money has been donated by people

chemotherapy
treatment of cancer with drugs

chloroquine
an antimalarial medicine

cholera
a life-threatening disease affecting the gut caused by *vibrio cholerae* bacteria

cirque
corrie (used in England and North America)

cirrus clouds
wispy high-level clouds that often herald the approach of a depression

climate change
term used to describe significant long-term changes in world weather patterns

clints
flat limestone blocks separated from each other by grykes on a limestone pavement

cold front
the boundary between cold and warm air where air is rapidly forced upwards creating cumulonimbus clouds, often giving short heavy downpours

column
where a stalactite and stalagmite have joined together; also known as a pillar

collision zone
where two continental plates are colliding, causing the crust to buckle and form fold mountains

Common Agricultural Policy
the European Union's system of farming programmes which give grants and subsidies to farmers

commuter
someone who travels some distance to and from their work each day

confluence
the point where two rivers join

congestion
when road traffic is so heavy that jams occur and journey times are lengthened

conservative plate boundary
where two of the Earth's plates are moving past each other, often sticking and causing earthquakes

constructive plate boundary
where two oceanic plates are moving apart, leading to the formation of new crust through volcanic activity

continentality
the effect that large land masses have on the weather

contour
a line drawn on a map joining points of equal height above sea level

contraception
birth control methods used to prevent pregnancy

conurbation
a large city that has spread outwards and joined up with other towns in the area (e.g. Clydeside, Greater Manchester)

corrasion
erosion caused by the scraping and scouring effect of material carried by rivers (or glaciers or waves)

corrosion
erosion caused by chemical reactions between rock and water

corrie
a deep, rounded hollow with steep slopes carved out of a mountain by glacial erosion.

corrie lochan
a lake in the saucer-shaped hollow at the bottom of a corrie

counter-urbanisation
when people move out of cities to get away from the problems associated with them, such as pollution and congestion

crater
the depression or hole at the top of a volcano from which eruptions happen

crop rotation
when different crops are planted in a field in successive years to help maintain soil fertility

cul-de-sac
a road that stops and doesn't lead anywhere else

cumulonimbus
towering clouds that often bring heavy downpours and sometimes electrical storms

cumulus
clouds with a fluffy appearance formed due to condensation caused by convection currents

curvilinear
pattern of roads consisting of many crescents and cul-de-sacs used in modern housing areas to create safer residential environments

cwm
corrie (used in Wales)

cyclone
a hurricane in the Indian Ocean

death rate
the number of deaths per 1000 people per year

Glossary

DEET
diethyl-meta-toluamide: a substance used in insect repellents

deforestation
the total clearance of forested land

delta
deposits found at the mouth of a river (often forming islands)

demographic transition model
a model showing how birth and death rates change over time

densely populated
an area with a lot of people per square kilometre

dependency ratio
a measurement of population which compares the working population (16–64 years) with the dependent population (0–15 and 65 years or over)

deposition
when material is left behind by rivers, ice or the sea

depression
an area of low pressure associated with active weather fronts

desertification
when land gradually turns into desert

destructive plate boundary
where two of the Earth's plates are moving towards each other and one is forced beneath the other and destroyed

developing country
usually quite a poor country with a low standard of living and a lack of services; also known as an economically less developed country (ELDC)

developed country
usually quite a wealthy country with a high standard of living and good services; also known as an economically more developed country (EMDC)

development gap
prices and wages are much higher in developed than in developing countries so for financial comparisons to be meaningful this difference must be taken into account

differential erosion
where soft rocks are eroded more quickly than adjacent hard rocks because they are less resistant

dormitory settlement
a place where a lot of people live but have to commute to work somewhere else

drainage basin
the land drained by a river and its tributaries; also referred to as the catchment area

drip tips
the pointed end of the leaves of rainforest plants which have evolved to shed water quickly

dripstone feature
deposits of calcite found underground in limestone areas

drought
prolonged period of dry weather

dry valley
a valley without a river flowing along it

eastings
the numbered grid lines along the bottom and top of a map that get higher towards the east

ecotourism
a type of holiday that is environmentally friendly and sustainable, usually involving small numbers of tourists

ELDC
economically less developed country; also known as a developing country

EMDC
economically more developed country; also known as a developed country

emergency aid
aid given quickly following a major disaster such as an earthquake

emergents
the tallest trees in the tropical rainforest (up to 50 metres high or more)

endemic disease
a disease that is restricted to certain areas or populations

environment
the surroundings in which people, plants and animals live

environmental stewardship
where the environment is looked after and managed in a way that is kind to plants and animals

epicentre
point directly above the focus of an earthquake where damage is usually worst

epidemic disease
an infectious disease that spreads quickly amongst the human population

epiphytes
plants which grow on others, especially in the tropical rainforest, deriving most of their nutrition from rain and the surrounding environment

erratic
a large boulder that has been transported and deposited by ice in an area where it is out of place

erosion
the wearing away of land caused by moving ice, rivers or waves

estuary
a section of a river which has sea water flowing in it

eutrophication
where excess nitrates and phosphates from farmland build up in water systems resulting in harmful environmental consequences

evergreen revolution
the use of new environmentally sustainable agricultural schemes in the developing world

exploitation
the selfish use of resources or people for economic gain without regard for the human or environmental consequences

explain
give reasons for

exports
goods and services sold to other countries

eyot
an island formed by deposition found in a river channel

fair trade
trade where the producers (usually in developing countries) get a fair price for their goods

falciparum malaria
the most deadly form of malaria

farm diversification
when farms branch out into new ways of getting income, apart from traditional food production

favela
shanty town in Brazil

favelados
residents of a favela

fetch
the distance over which waves build up

firn
hard compacted snow; the stage between freshly fallen snow and the formation of glacial ice (also known as névé)

flood plain
the area of flat land on either side of a river that is flooded when the river bursts its banks

focus
the point beneath the ground where an earthquake has taken place

fodder
crops grown to feed to livestock

fold mountains
mountain range formed by the buckling of the Earth's crust where two continental plates are colliding

food miles
the total distance which food travels between its place of origin and the point of consumption

fossil fuels
energy resources such as coal, oil and natural gas formed from the fossilised remains of plants and animals

freeze-thaw action
see frost shattering

frost shattering
a type of weathering where rock is broken up by repeated freezing and thawing of water contained in small cracks (also known as freeze–thaw action)

fumarole
vent emitting sulphurous gases in a volcanic area

function
the main purpose of a town or city; settlements often have several different functions, such as port, market town, administrative centre

genetically modified crops
where the DNA of crops has been altered or engineered to make them grow faster or to enable them to survive in cooler, drier conditions, for example

geo
a very narrow and steep-sided inlet from the cliff edge

geothermal power
energy generated from naturally occurring heat found in rock beneath the surface, much of which originates from the Earth's molten interior

glacier
very slow moving mass of ice flowing down a valley from high mountain corries

glaciokarst
limestone scenery that was first exposed by the scouring action of glaciers and ice sheets

globalisation
the tendency for different parts of the planet to become increasingly integrated through trade, rapid transport and the spread of ideas and values via the internet and social media

global warming
the gradual rise in world temperatures thought to be due to the increase in levels of carbon dioxide and other gases due to human activities

GM
genetically modified (used in relation to crops); see above

gorge
a very steep-sided valley caused by the retreat of a waterfall over time; in limestone areas gorges may be formed by fast flowing glacial meltwater

grant
money given to an industry such as farming by the government or the European Union to help its development

green belt
land around the edge of a city where building is severely restricted in order to protect the countryside and stop the outward spread of built-up areas

greenfield site
land which has never been built on before, often at the edge of a town or city

greenhouse effect
natural process by which gases in the atmosphere act like a blanket, trapping the sun's heat

greenhouse gas
a gas which helps to trap heat in the Earth's atmosphere (e.g. carbon dioxide)

Greenpeace
a well-known environmental campaign group

green revolution
use of a variety of new farming techniques during the latter half of the 20th century designed to increase agricultural output

GLOSSARY

grid-iron pattern
parallel pattern of roads intersected at right angles by another series of parallel roads; often found in old inner city areas; can be referred to as rectilinear

grykes
enlarged joints between blocks of limestone known as clints

haar
coastal fog often found along Scottish North Sea coasts, caused by cool air from the sea blowing on to warmer land

hanging valley
small U-shaped valley left high above a larger U-shaped valley formed by a tributary glacier that could not erode down as far as the glacier in the main valley

HDI
Human Development Index: a figure showing a country's level of development, calculated by combining social and economic indicators: these are average income, life expectancy and average number of years at school

headland
a cliff protruding out to sea; usually the highest point for some distance

heavy industry
factories producing manufactured goods that require large amounts of heavy-weight raw materials, e.g. shipbuilding, steel making

HEP
hydro-electric power: electricity generated by the force of water falling onto turbines

high-order service
shop or facility where expensive products are sold but purchased only occasionally by each individual

HIV
human immunodeficiency virus: causes infection that can lead to AIDS

hot desert
hot dry areas usually found close to the tropics which have less than 250 millimetres of rainfall per year

hurricane
a severe tropical storm where steady wind speeds exceed 120 kilometres per hour

hydraulic action
type of erosion caused by the sheer force of water or waves breaking off material from the bed or banks of a river or from the coastline

hydroponics
the science of growing crops without soil

HYV
high yielding variety (of crop)

ice plucking
where ice at the bottom of a glacier freezes onto weathered rock and pulls it away as the glacier moves forward

ice sheet
where many glaciers have joined together to form a cloak of ice that completely covers the land to a depth of hundreds of metres

immigrant
a person arriving in a country to live or work

impermeable
does not allow water to pass through

imports
goods and services bought by a country from abroad

indigenous people
the communities first associated with living in a particular area (e.g. the Inuit people in the tundra)

industrial estate
a group of factories in a specially built development, designed to give them all the services and space they need

infant mortality
the number of deaths of children under one year old per 1000 live births

informal economy
trade which is not recorded, for example where goods are exchanged without money changing hands and no records are kept

infrastructure
the network of roads, telecommunications and utilities, such as water, electricity, gas and sewerage, needed to allow the development of an area to take place

inner city
land-use zone immediately beyond the CBD where factories and workers' houses were built during the industrial revolution

insolation
energy from the Sun that arrives in the Earth's atmosphere

intensive farming
where maximum use is made of every scrap of agricultural land, often involving controversial practices such as heavy use of chemical fertilisers and pesticides

interlocking spurs
hill slopes projecting into a V-shaped valley between which a river meanders

intermittent drainage
where streams disappear underground and then reappear as they flow from impermeable to permeable rocks and back again

irrigation
the application of water to farmland to supplement natural rainfall

isobar
a line on a synoptic chart joining points of equal air pressure

joint
a vertical crack in the rock

karst scenery
an area of limestone scenery named after an extensive area of limestone in Slovenia

kayak
a sea-going canoe originally developed by the Inuit people to hunt seals in the Arctic and tundra regions

kwashiorkor
a diet-deficiency disease caused by lack of protein

lagoon
the area of water behind a sand bar that is now cut off from the sea

lahars
a deadly avalanche of mud, water, lava and ash that can flow very quickly down a river valley following a volcanic eruption

land degradation
when the natural environment is damaged, often by harmful human activities

land reform
where ownership of land is redistributed

land-use conflict
disagreement over how an area of land should be used

lateral moraine
rock debris deposited at the side of a glacier, often along the length of a valley

latitude
how far north or south a place is from the equator; measured in degrees

lava
molten rock that erupts on to the Earth's surface

leaching
when minerals are washed out of the soil by heavy rain

levées
natural banks of deposits that build up above the bank on either side of a river

lianas
vine-like plants that wind their way round others in the tropical rainforest

life expectancy
the average number of years that a person born in a particular country is expected to live for

light industry
factories producing goods with a high value to weight ratio, e.g. computers, electrical equipment

limestone pavement
a naturally flat expanse of limestone blocks or clints separated by grykes

limestone scar
a long, sometimes vertical, exposure of limestone on a steep slope created by the scouring effect of a glacier

literacy rate
the proportion of people who can read and write

load
the material carried by a river

longshore drift
the movement of material along the coast due to the action of the sea

long-term aid
help given to a country over many years, often for an HEP scheme or rebuilding after a major disaster

low-order service
shop or facility where low-cost products are sold and purchased frequently by each individual

magma
molten rock beneath the Earth's surface

magma chamber
areas beneath a volcano where molten rock collects

malaria
life-threatening parasitic disease transmitted by the anopheles mosquito

malathion
an insecticide used against mosquitoes in order to reduce the spread of malaria

market town
town surrounded by a large number of farms where agricultural products are brought to be sold

meander
a bend in a river

medial moraine
line of rock debris down the centre of a valley created by the lateral moraines of two tributary glaciers joining together

mefloquine
an antimalarial medicine

migration
the movement of people from one place to another to live or work

Milankovitch theory
a collective term describing possible changes to the Earth's tilt and orbit which could explain global changes in climate

millibars
units of measurement for air pressure

millimetres
one tenth of a centimetre; units of measurement for precipitation

mixed farming
where both crops and animals are produced on a farm

moraine
material deposited by a glacier

mouth
the point where a river flows into the sea

multinational company
a business with its headquarters in one (usually developed) country but bases and/or factories in many other countries

multiplier effect
when the arrival of one large company in an area attracts in many other companies

muskeg
boggy land found during the summer in tundra areas where melted snow cannot drain away because of permafrost

Glossary

National park
a large area of attractive scenery that is protected from harmful developments so that people living in and visiting the area can enjoy a relatively unspoilt environment

natural increase
how fast a country's population is increasing; calculated by finding the difference between a country's birth and death rates; often referred to as population growth rate

névé
hard compacted snow; the stage between freshly fallen snow and the formation of glacial ice (also known as firn)

North Atlantic Drift
a warm current that flows in a north-easterly direction across the North Atlantic, affecting the climate in north-west Europe

northings
the numbered grid lines up the side of the map that get higher towards the north

occluded front
where warm and cold fronts meet, often causing bad weather

okta
a measurement of cloud cover representing one eighth of the visible sky

ORS
oral rehydration solution: a mixture of salts, glucose and water used in the treatment of cholera

ORT
oral rehydration therapy: cholera treatment using ORS

organic farming
agriculture where the inputs are mostly natural (e.g. only natural, instead of chemical, fertilisers are used)

outwash plain
area beyond the snout of a glacier that is covered in meltwater deposits

overcultivation
where soil becomes exhausted due to farming activities that cannot be sustained; one of the causes of desertification

overfishing
where fish stocks become low and may be threatened with extinction because too many fish are being caught

overgrazing
where livestock are allowed to graze on marginal land for too long resulting in vegetation being destroyed; one of the causes of desertification

overpopulation
where the resources of an area can no longer support the number of people living there; one of the causes of desertification

ox-bow lake
an abandoned meander that is crescent shaped and now separate from the main river channel

parasitic cone
secondary vent and crater on the side of a bigger volcano

park and ride
scheme designed to reduce congestion, where people can park their cars for free at the edge of the city and take a bus into the centre

pastoral
farming that involves only the rearing of animals (e.g. sheep)

pedestrianisation
where traffic is banned from a street to create a safer and more pleasant environment for shoppers

permafrost
permanently frozen sub-soil found in tundra areas

permeable
allows water to pass through

phreatic cave
a cave formed mainly beneath the water table, leaving smooth sides and a tubular shape

pillar
where a stalactite and stalagmite have joined together; also known as a column

pneumonia
inflammation of one or both lungs usually caused by an infection

polar continental
air mass originating over land areas in northern Europe bringing bitterly cold weather

polar maritime
air mass originating in northerly sea areas bringing cold wet weather

pollution
contamination of the environment by noise, dirt, fumes or other harmful substances due to human activities

population density
the average number of people living in each square kilometre of an area

population growth rate
how fast a country's population is increasing; calculated by finding the difference between a country's birth and death rates; often referred to as natural increase

population pyramid
a graph showing the proportion of males and females in each age group for an area or country

population structure
how a population is made up according to gender and age

PQLI
Physical Quality of Life Index: a figure to measure a country's level of development, which combines its life expectancy, literacy and infant mortality rates

precipitation
different types of moisture coming mainly from clouds landing on the ground; includes rain, snow, sleet, hail and drizzle

precision farming
the use of satellite technology and exact soil measurements to allow computerised application of fertilisers to different areas of a field to help maximise crop yields

prevailing wind
the most common wind direction in a particular area

primary industry
an industry that extracts raw materials from the land or sea; includes farming, fishing, mining, quarrying etc.

push factors
things that might make people want to leave an area, e.g. lack of jobs, drought

pull factors
things that might attract people to an area, e.g. better standard of living, higher pay

pyramidal peak
a jagged or pointed mountain that has been affected by the formation of corries on three or more sides

pyroclastic flow
a lethal mix of hot rocks, volcanic ash and gas that travels very quickly down the side of a volcano due to gravity

quality of life
a measure of how comfortable and content people are with their lives

quinine
a substance occurring naturally in the bark of the cinchona tree that has been used as the basis of some antimalarial drugs such as chloroquine

quotas
where a limit has been imposed on a commodity, e.g. on the production of milk or on the number of fish caught

range of temperature
the difference between the highest and lowest average temperatures

raw materials
natural resources used to make other products

rectilinear
parallel pattern of roads intersected at right angles by another series of parallel roads; often found in old inner city areas; can be referred to as a grid-iron pattern

refugee
someone who has been displaced from their home area by war, persecution or a natural disaster

relief
the shape and height of the land

recurved spit
a sand spit that has secondary spits towards its end pointing inland due to temporary shifts in wind direction

renewable resource
a resource such as wave, wind or hydro-electric power that can be used over and over again without it ever running out

resurgence
the point where a stream reappears on the surface from beneath the ground, especially where limestone meets impermeable rock

ribbon loch
a long narrow loch formed in a U-shaped valley due to glacial erosion

Richter scale
the logarithmic scale used to measure the strength of earthquakes

river beach
a beach on the inside of a meander formed by the deposition of river material

river braiding
where a river deposits material in the middle of its channel, creating islands and causing it to split temporarily into different channels

river cliff
a vertical or overhanging slope on the outside of a meander caused by river erosion

route centre
a settlement where many different roads (or railways) meet; this might have been one of the reasons for its growth

RSPB
Royal Society for the Protection of Birds

rural
to do with the countryside

rural-urban fringe
land at the very edge of the city where the built-up area meets the countryside

rural-urban migration
movement of people from the countryside to the city, usually in an attempt to improve their standard of living

rurban fringe
abbreviation of rural-urban fringe (see above)

Saffir-Simpson scale
a classification of hurricanes according to the strength of their sustained wind speeds

salt marsh
found behind a sand spit where salt-tolerant plants have started to grow in shallow sea water, trapping yet more sand

sand bar
a sand spit that stretches right across a bay and is joined to the land at both ends

sand spit
a narrow finger of land protruding out into the sea, created by longshore drift and the deposition of sand

science park
a type of industrial estate where the main companies involved are high-technology research and design activities, e.g. biotechnology

scree slope
a pile of broken, frost-shattered rock built up from the bottom of a slope

sea arch
where a cave extends right through a headland

GLOSSARY

sea stack
a pillar of rock standing up out of the sea, caused by the roof of an arch collapsing

sedimentary rock
rock that has been formed due to deposition of fine material on an ancient sea bed

self-help scheme
a project, usually in a developing country, where for a small amount of money local people are able to learn new skills and help their own communities in a variety of different ways

set-aside land
land that is left uncultivated and for which the farmer receives a government grant

settlement
a place where people live

shake hole
a depression, several metres deep, found in an area of limestone, often filled with boulder clay

shanty town
a ramshackle group of houses, often at the edge of developing world cities, which the inhabitants have built themselves using whatever materials they can get hold of

shifting cultivation
type of farming carried out by indigenous people in tropical rainforests, where farming activities move on to new ground every few years due to poor soil fertility

short-term aid
aid given usually to developing countries that is intended to help out immediately following a crisis such as an earthquake; can include tents, blankets, clean water, medical supplies etc.; often known also as emergency aid

silt
fertile material left behind after a river floods (also called alluvium)

site
the land on which is a settlement is built

situation
where a settlement is in relation to other places in that area; also called location

snout
the front end of a glacier; sometimes referred to as the toe

socio-economic indicator
information about the population and wealth of countries which allows meaningful comparisons to be made

soluble
can be dissolved in a liquid

source
the point (usually in the hills) where a river begins

sparsely populated
an area with very few people per square kilometre

sphere of influence
the area around a settlement or service from which it draws shoppers/customers

stalactite
a deposit of calcite hanging from the roof of a cavern

stalagmite
a deposit of calcite built up from the floor of a cave

standard of living
how well off a person or population is

storm surge
a temporary increase in sea levels caused by intense low pressure and strong winds which can lead to devastating coastal flooding

stratus
low-layer cloud, often found close to a warm front

striations
scratches left on flat rock that have been caused by rocks frozen into the base of an ice sheet or glacier, abrading the bed rock as the ice moved along

stump
the remains of a collapsed stack

subsistence
used to describe a self-sufficient lifestyle or type of farming that supplies just enough for a person's needs and no more

suburbs
the outlying districts of a city made up mainly of housing areas

subduction zone
an area where a plate is being forced down into the mantle and destroyed

sustainable
something that can be maintained indefinitely without damaging the environment or depleting resources for future generations

swallow hole
where a stream disappears underground down an enlarged joint

swash
water from a breaking wave moving up the beach

synoptic chart
weather map using symbols for fronts, isobars and different weather stations

tariff
a tax imposed selectively on goods being brought into a country that is designed to protect local producers of the same product

tenement housing
a row of buildings containing flats joined together, usually three or four stories high and with a common entrance close and stairwell, built in Scottish inner city areas

terraced housing
a row of houses joined together, usually just two stories high and each with their own entrance on to the street, found in inner city areas in England

terminal moraine
mound of rock debris deposited by a glacier at its snout

tertiary industry
industry providing services to other people, e.g. transport, medical and retail services

tidal limit
the furthest point up a river which the tide regularly reaches

tombolo
a sand spit joining an offshore island to the mainland

tourist resort
a town or place that caters for large numbers of tourists

trade
the movement of goods and services between countries

trade alliance
a group of countries which have joined together to give themselves more economic power in world trade, e.g. the European Union

trade deficit
where the value of a country's imports exceeds the value of its exports, resulting in an overall loss

trade surplus
where the value of a country's exports exceeds the value of its imports, resulting in an overall profit

transportation
the movement of material by rivers, ice or the sea

tributary
a small river that joins a larger one

tropical storm
an intense area of low pressure with sustained wind speeds in excess of 60 kilometres per hour, formed due to warm oceanic surface temperatures

tropical hardwood
timber taken from tropical rainforests such as teak, ebony, mahogany and rosewood

tropical continental
air mass originating from land near the tropics that brings warm dry weather

tropical maritime
air mass originating from sea areas over the Tropics that brings warm wet weather

truncated spur
a slope on a hillside that originally extended further into a valley but which has had its end sliced off by a glacier

tundra
cold desert areas found in extreme northern latitudes

typhoon
a hurricane in the western Pacific Ocean affecting south-east Asia

urban
to do with built-up areas such as towns and cities

urbanisation
the increase in the proportion of people living in towns and cities

urban redevelopment
the replacement of derelict buildings, old tenements or terraced housing often with high-rise blocks in inner city areas

urban sprawl
the relentless outward growth of cities into the surrounding countryside

U-shaped valley
a valley with steep sides and a flat bottom that has been carved by a glacier, giving the valley a U-shaped cross-sectional profile

vadose cave
a cave formed mainly above the water table that is less regular in shape than a phreatic cave as all surfaces are not in contact with water

vent
in a volcano this is the pipe up which molten rock travels to erupt on to the surface

vertical farming
indoor farming (hydroponics) in multi-storey buildings

volcanic bomb
a lump of semi-molten rock that can be catapulted a great distance during an eruption

volcanic plug
solidified lava blocking the crater or vent of a volcano

voluntary aid
aid given by charitable organisations, such as Oxfam and the Red Cross, which are funded by donations mainly from the public; also known as charitable aid

V-shaped valley
a valley with steep sides and a narrow bottom that has been carved by a river, usually in its upper course, giving the valley a V-shaped cross-sectional profile

warm front
the boundary between cold and warm air where the warm air rises slowly up over the cold air causing the formation of stratus and cumulus clouds, giving a prolonged spell of rain

water sink
the point where a stream disappears beneath the ground by soaking through a permeable rock such as limestone

wave cut notch
the most eroded part of a cliff at its base, where waves do the most damage

wave cut platform
an area of flat rock stretching from the base of a cliff out to below the water's edge; this is all that remains after cliffs have been eroded away altogether

wave pounding
where the weight of water hitting a cliff face repeatedly erodes it

weather element
a particular atmospheric condition, such as temperature, which, together with other elements, make up the weather

weather front
the boundary between air masses of different temperatures

weathering
the breakdown of rocks by physical and chemical processes caused by the weather

yield
in relation to farming, this is the amount of crops obtained from each hectare of land